HAWAI'I THE FIRES OF LIFE

Rebirth in Volcano Land

Five Decades of Vegetation Development in the Devastation Area, Site of the 1959 Kīlauea Iki Eruption, Hawai'i Volcanoes National Park

Other book titles on which Dieter Mueller-Dombois is the principal author:

Aims & Methods of Vegetation Ecology (with H. Ellenberg) John Wiley & Sons 1974, republished as classical textbook by The Blackburn Press 2002.

Island Ecosystems: Biological Organization in Hawaiian Communities (ed. with K.W. Bridges and H.L.Carson) Hutchinson Ross Publishing Company 1981.

Vegetation of the Tropical Pacific Islands (with F.R.Fosberg) Springer-Verlag 1998.

Biodiversity Assessment of Tropical Island Ecosystems: PABITRA Manual for Interactive Ecology and Management (ed. with K.W. Bridges and Curtis Daehler) published online www.botany.hawaii.edu/pabitra 2005.

HAWAI'I THE FIRES OF LIFE

Rebirth in Volcano Land

Five Decades of Vegetation Development in the Devastation Area, Site of the 1959 Kīlauea Iki Eruption, Hawai'i Volcanoes National Park

Garrett A. Smathers
Dieter Mueller-Dombois

Copyright © 2007 by Mutual Publishing, LLC.

All Rights Reserved. No part of this book may be reproduced in any form or by any electronic or mechanical means, including information storage and retrieval devices or systems without prior written permission from the publisher, except that brief passages may be quoted for reviews.

Library of Congress Cataloging-in-Publication Data

Smathers, Garrett A.
    Hawai'i : the fires of life : rebirth in volcano land, five decades of vegetation development in the devastation area, site of the 1959 Kilauea Iki eruption, Hawai'i Volcanoes National Park / Garrett A. Smathers, Dieter Mueller-Dombois.
       p. cm.
    Summary: "A scientific chronicle based on 50 years of ecological research and study of the rebirth of vegetation in the devastation area of Kilauea Iki on Hawaii's Big Island"--Provided by publisher.
    Includes bibliographical references.
    ISBN-13: 978-1-56647-818-2 (softcover : alk. paper)
    ISBN-10: 1-56647-818-9 (softcover : alk. paper)
    1. Plant ecophysiology--Hawaii--Kilauea Volcano Region. 2. Plant succession--Hawaii--Kilauea Volcano Region. 3. Plants--Effect of volcanic eruptions on--Hawaii--Kilauea Volcano Region. 4. Kilauea Volcano (Hawaii) I. Mueller-Dombois, Dieter, 1925- II. Title.
    QK473.H4S63 2006
    581.9969'1--dc22
                    2006037128

ISBN-10: 1-56647-818-9
ISBN-13: 978-1-56647-818-2

Design by Mark Nakamura
Cover Design and Production by Emily R. Lee

First Printing, April 2007
1 2 3 4 5 6 7 8 9

Mutual Publishing, LLC
1215 Center Street, Suite 210
Honolulu, Hawai'i 96816
Ph: (808) 732-1709
Fax: (808) 734-4094
e-mail: info@mutualpublishing.com
www.mutualpublishing.com

Printed in Korea

Contents

Preface vii
Acknowledgements ix
The Devastation Area 1
Important Background
   Brief History 7
   Clarification of Terms 10
Research Protocol
   A Unique Opportunity 13
   Vegetation Surveys 18
Vegetation Development
   Habitat 1: The Crater Floor Habitat 21
   Habitat 2: The Cinder Cone Habitat 33
   Habitat 3: The Spatter Habitat 45
   Habitat 4: The Tree Snag Habitat 59
   Habitat 5: The Survival Tree Habitat 79
   Habitat 6: The Thin Fallout Area 97
A New Lava Flow 113
Summary and Conclusions 117
Checklist of Plants 125
Photo Credits 132
Figure Credits 132
Bibliography 133
Index 137
About the Authors 142

Preface

The story begins with the eruption of Kīlauea Iki in November 1959. Garrett A. Smathers, a Park Naturalist with the National Park Service, was transferred to Hawai'i Volcanoes National Park shortly after the eruption. He became curious about how such destruction of vegetation would recover, and what role native versus nonnative plants would play in the recovery process of early primary succession. The second author, Dieter Mueller-Dombois, became a member of the Botany Faculty at the University of Hawai'i in 1963. Soon thereafter the authors met, and both agreed that long-term ecological research (LTER) of vegetation development in the Devastation Area would be of value for science and application.

Most studies of vegetation development and succession are based on substituting space for time. This means that contemporary vegetation types, which developed after different known dates of major disturbances in the same climatic zone, are placed into a chronosequence to explain how they have formed and developed over time. The space-for-time substitution studies always contain a hypothetical element in that they assume bioenvironmental conditions to have remained constant over the chronosequence.

Rather few studies are based on long-term observations of the same place. Such *in situ* succession studies are few, because they depend on the opportunity for LTER and the documentation of real change over time in the same area.

Such documentation is herewith provided in popular form of photographs with some explanation of the patterns and processes of vegetation development over a time frame of 46 years. In terms of primary succession, this is only a small window of time, since vegetation development to closed mature forest is estimated to take at least 200 years in Hawai'i's montane rain forest environment and still longer in the seasonal environment.

We have learned that each of the different habitats supplied with a new volcanic surface has a unique sequence of early vegetation development. This process is now complemented and accelerated by nonnative species. But a core group of robust native pioneer plants still dominate the colonizing process. While the initial invasion and recovery processes are unique in relation to substrate and habitat type, we can expect a conversion of vegetation development to greater uniformity in the respective climatic zones only in the later stages of primary succession.

Forests will definitely develop in both the rain forest and seasonal forest zones across the Devastation Area according to the climatic site potential. These forests will also still be dominated by the 'ōhi'a lehua tree (with some help by Park Resources Management), unless such development to forest maturity is shortened by another major volcanic disturbance in the same area. The latter is rather likely. It would render the Devastation Area to remain a pioneer vegetation zone as it is today, and native pioneer plants will also remain the dominant initial colonizers, because—as the reader will see—most nonnative invaders depend on the native colonizers for access to invade new volcanic surfaces in this terrain.

Acknowledgements

Our LTER project received a great deal of help over the years. Park Rangers Robert T. Haugen and Stewart Branson assisted during the early stages in setting up photo stations together with Jerry P. Eaton and Donald H. Richter from the Volcano Observatory. Ranger Haugen made the first photo coverage. Park Superintendents Fred Johnson and Daniel J. Tobin, Jr. gave the Devastation Area a special status for research and protection. Park Naturalists, who took a keen interest in the project as a special area for interpretation, included Robert L. Barrel, William W. Dunmire, Richard S. Rayner, Richard MacBride, and Dwight Hamilton. Financial support was initially provided by grants from the Hawai'i Natural History Association, the USDI National Park Service and later in part by NSF grants (GB 4686, IBP 23230, BSR 18526) to the co-author.

The bluegreen algae were identified by specialists, originally by Francis Drouet and recently for the many name corrections by Allison Sherwood. The lichens were identified by I. MacKenzie Lamb and Mason E. Hale. Bryophytes were originally identified with the help of William Hoe and recent name changes were updated by Mashuri Waite. Technical assistance in the periodic resurveys was given by Ranjit G. Coorey and Nengah Wirawan in 1974, by Grant Gerrish and R. Lani Stemmermann in 1981, by Cynthia C. Lipp and Donald D. Drake in 1988, and by Julia Williams in 1998. Others providing help in special subprojects included Robert A. Wright and Louis D. Whiteaker. Park Service Personnel and Researchers of the USDI Kīlauea Field Station who supported this LTER in the later phases included Dan Taylor, Tim Tunison, Darcy Hu, Rhonda Loh, and Linda Pratt. Linda, in particular, gave expert botanical advice.

Computer print formatting and design help was provided expertly by Mark Nakamura from the UH Mānoa Center for Instructional Support. Computer processing and editorial help was given by Annette Mueller-Dombois. To all these persons, we owe our deep felt thanks and appreciation. Their sustained interest and support made this publication possible.

The manuscript was peer reviewed by three biologists familiar with the area, Paul Banko, Rhonda Loh, and Linda Pratt, and by two geologists, Jack Lockwood and Don Swanson, who are familiar with the volcanology of the area. All provided constructive comments. We owe them our special thanks.

Garrett A. Smathers, PhD
Former Chief Scientist
NPS Science Center
NASA–National Space
Technology Laboratories
Bay St. Louis, Mississippi

Dieter Mueller-Dombois, PhD
Emeritus Professor, Dr.h.c.
Department of Botany
University of Hawai'i
at Mānoa
Honolulu, Hawai'i

The Devastation Area

On November 14, 1959, Kīlauea Iki, a small crater outside the east rim of Kīlauea Volcano's caldera, erupted and spewed lava materials into the air in form of volcanic ash. At the same time lava was flowing onto the crater floor burying a former rain forest. The eruption lasted 36 days.

Lava of over 2000°F (1093°C) flowed from the base of the fountain onto the crater floor, creating a lava lake approximately 400 feet (122 m) deep. At one time the fountain reached a height of 1900 feet (580 m), the highest ever recorded in Hawai'i (see Photo 2).

PHOTO 1. Hawai'i Volcanoes National Park; home of Kīlauea—one of the world's most active volcanos.

Photo 2. 1959 Kīlauea Iki lava fountain.

The Devastation Area 3

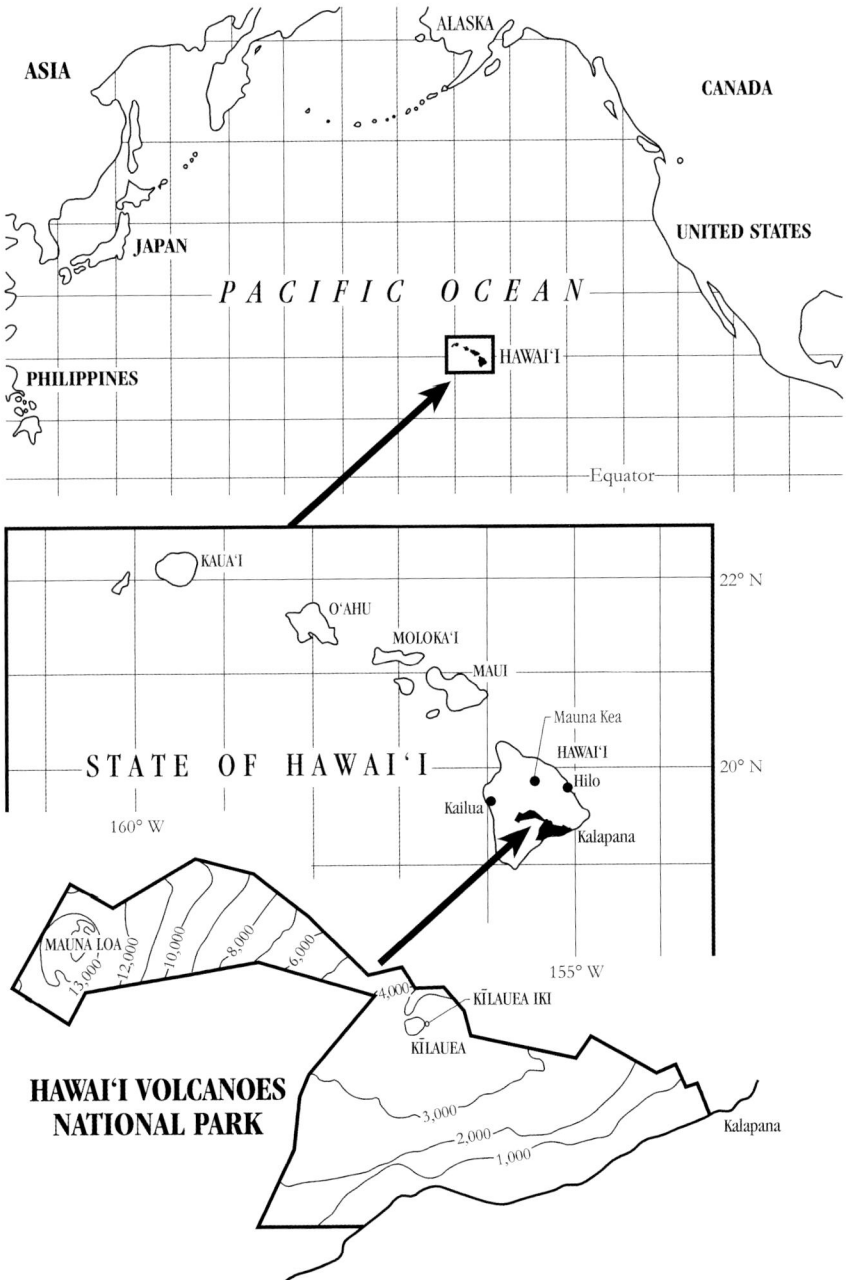

FIGURE 1. Hawai'i location map with Kīlauea Iki eruption site. Contour lines in feet. The Park boundary here was drawn prior to the extension of 'Ōla'a Tract and Kahuku Ranch.

Kīlauea Iki and Keanakākoʻi are pit craters east of the summit caldera of Kīlauea Volcano on the Island of Hawaiʻi. Halemaʻumaʻu is the main vent of Kīlauea Volcano.

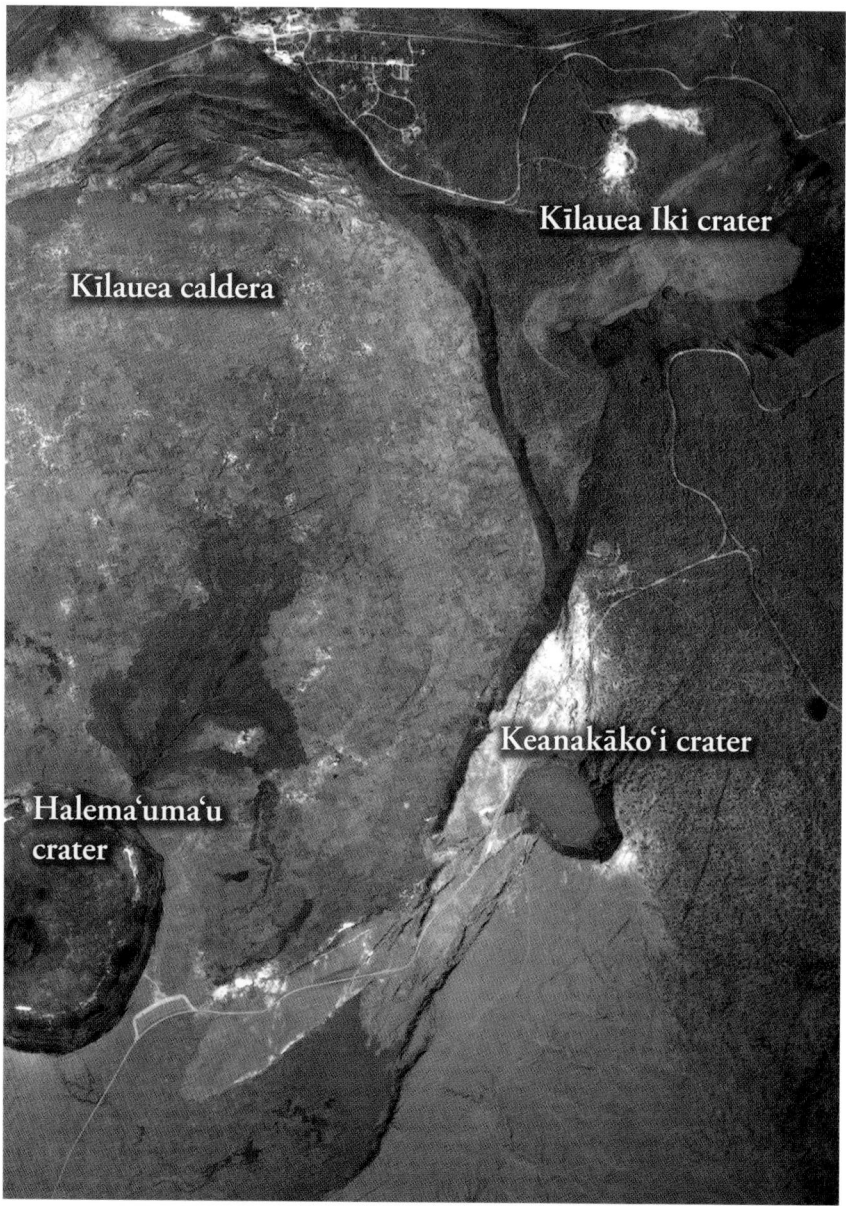

PHOTO 3. Aerial view of Kīlauea Volcano summit prior to 1959 Kīlauea Iki eruption.

The Devastation Area

The prevailing northeast trade wind spread cinder-ash and spatter around the crater and light ash material over 4 miles (6.4 km) on the leeward southwestern side of the Kīlauea volcano summit area.

PHOTO 4. The November 1959 Kīlauea Iki eruption began with a line of fountains along the crater wall.

PHOTO 5. A lone ʻōhiʻa lehua tree snag protrudes with its top through a 40-feet (12 m) blanket of cinder-ash. In the background is the 152-feet (46 m) high Puʻu Puaʻi cinder cone created by airfall.

A blanket of pyroclastic tephra or cinder-ash (fire-broken rock deposited from airfall), varying from 200 feet (61 m) deep at the fountain base to 1 inch (2.5 cm) deep in the leeward Kaʻū Desert, destroyed or damaged part of a native ʻōhiʻa lehua (*Metrosideros polymorpha*) seasonal dry forest on the east side of the summit of Kīlauea Volcano. A section of native montane tropical rain forest on Kīlauea Iki crater floor was covered with a lake of lava and another section was buried under the cinder cone. The entire area affected was named **The Devastation Area** (see map Figure 2, page 15).

Important Background
Brief History

The study of the Devastation Area in Hawaiʻi Volcanoes National Park as initiated in 1960, became the first National Park Service supported LTER (long-term ecological research) program in the Hawaiian National Parks. In 1966 the Devastation Area Study helped secure funding through the Hawaiʻi Natural History Association for the preparation of an "Atlas for Bioecology Studies in Hawaiʻi Volcanoes National Park," a basic background document, produced by the University of Hawaiʻi Botany Department. By 1967 the Devastation Area Study was sharing laboratory and office space in a Cooperative Park Studies Unit (CPSU) with the U. S. Fish and Wildlife's Endangered Hawaiian Bird Research Program and the University of Hawaiʻi at Mānoa.

The above mentioned Cooperative Park Research Programs attracted the IBP (International Biological Program) studies from 1971–1976 to the Park for conducting a major part of its Island Ecosystems Research there. In 1975 the CPSU became institutionalized with funding from the Park Service in the Botany Department at the University of Hawaiʻi at Mānoa, where it is still operating as a major research unit today. Recently, the unit was renamed PCSU (Pacific Cooperative Studies Unit and more recently still CESU for Cooperative Ecosystem Studies Unit) to emphasize its expanded tasks into the Pacific U. S. Territories.

The Hawaiʻi IBP and the CPSU convinced park officials that the bioecology and evolutionary biology of this National Park ranks equal in importance to the Park's geology and volcanology in science and application. It also led to the realization that both the bioecological and geological processes provide for the unique nature of the Park's ecosystems.

The first decade of the Devastation Area Study was published by the U.S. Government Printing Office with the title "Invasion and Recovery of Vegetation after a Volcanic Eruption in Hawai'i," National Park Service (NPS) Scientific Monograph Series, Number 5, 1974.

This monograph was among the first basic scientific studies, initiated and published by the National Park Service itself. It provided park managers reliable information for applied and operational research on how to protect, manage and interpret the Hawaiian Park Ecosystems' unique natural resources. The protection of the natural and cultural heritage is now an interrelated task of the Park's resource management unit. Protection management offers continuing challenges because of the Park's highly dynamic ecosystems.

The Devastation Area Study and the IBP work in the Park were recognized by UNESCO (the United Nations Educational Scientific and Cultural Organization) and its MAB Program on Man and the Biosphere, as part of the International Network of Biosphere Reserves.

PHOTO 6. Hawai'i Volcanoes National Park recognized as a part of the International Network of Biosphere Reserves, November 20, 1980.

Important Background

The MAB designation, and continued scientific publications on the Hawai'i Volcanoes Devastation Area and IBP studies, caught the attention of the worldwide scientific and conservation

Photo 7. Hawai'i Volcanoes National Park designated as a World Heritage Site December 11, 1987.

Photo 8. Hawai'i Volcanoes National Park Research Center.

community. In 1987, the Park was designated as a World Heritage Site, because of its outstanding natural and cultural resources that form the common heritage of all mankind.

In time, park managers realized that larger laboratory and office space would be required, to house the growing cooperative multidisciplinary agency and university consortium personnel. To meet this need a Research Center was established in Hawai'i Volcanoes National Park. The Research Center (now officially named the Pacific Island Ecosystems Research Center, Kīlauea Field Station of the USDI Geological Survey) provides research needs to all the Hawaiian National Park Service areas and beyond. It is staffed with National Park Service research-resources management personnel, and houses personnel of the USGS (U.S. Geological Survey), BRD (Biological Research Division), and PIERC (Pacific Island Ecosystems Research Center). Also, research scientists from the University of Hawai'i, Stanford University, the B.P. Bishop Museum, and other research institutions work there on a transient basis.

Clarification of Terms

The term **vegetation** refers to the plant cover of an area. Vegetation is composed of species, populations and individual plants that share an area or habitat in various combinations. Some combinations may be recognized as plant communities, in particular when species occur together repeatedly in similar combinations or when they occupy a unique habitat. Usually, the plant cover or vegetation of an area can be subdivided into communities by dominant species, which are those that are most prevalent.

The term **flora** refers to a list of plant species and books that describe them. Such books are used to identify plant species by keys, drawings, and detailed technical descriptions. The term flora must not be confused with vegetation even though both may refer to the same area.

The term **habitat** refers to an area, place or site with all its environmental factors and constraints. The environmental factors include the essentials for plant life, such as water, nutrients, heat, and light. These factors are the physical and chemical components of climate and soil or substrate.

The term **vegetation development** is used in the following sections to illustrate what happened on each of the six habitats mapped in this study. Vegetation development is a process that may start with a disturbance. It may involve plant recovery and invasion of new plant life, establishment and reproduction, natural selection by habitat constraints, displacement in competition or self-decline of certain species in the course of succession, and eventually adjustment during maturation of the plant cover to the prevailing climate and soil conditions.

We can express this process in form of a six factor function (f) in an integral formula as:

Vegetation development = f (g, cl, d, fl, ac, e)

Where:
- g = ground condition, geoposition, geology, geomorphology
- cl = climate, regional, local, and microclimate
- d = disturbance factors including disturbance regime (repeated patterns)
- fl = flora of a region or area (as described in books or plant checklists)
- ac = accessibility of a plant species to reach the habitat in question
- e = ecological properties of species (survival and growth characteristics under the habitat constraints, plant forms and functions)

Time and space must be considered as two additional dimensions. **Time** is the major overriding dimension in this LTER project. **Space** is a major overriding dimension in relation to the size of a disturbed area. If all things are equal, a large area takes much longer to recover than a small area, primarily because of plant dispersal limitations.

Ecosystems can be understood as combinations of vegetation with their specific habitats, since vegetation is the major biological component of almost any natural terrestrial ecological system. However, in reality other organisms also form essential parts of an ecosystem. They include animals as seed dispersers, as pollinators, as ground disturbers, as consumers, and other organisms as ground dwellers, decomposers, parasites and saprophytes. These organisms together with the vegetation and their habitats form a functional system, an ecosystem. We can analyze these ecosystem components separately, as is done by different specialists, but to understand their dynamic behavior, we have to put these components together. This is achieved by synthesizing their relationships and appraising ecosystems holistically.

Landscapes are distinct areas of land that may be occupied or unoccupied with natural vegetation, forest plantations, agricultural fields, and/or human settlements, including cities and a combination of those. Landscapes are basic geological landforms together with built-up environmental features. They are also understood as land areas supporting several individual ecosystems or habitats as well as geographically larger ecosystems.

Research Protocol
A Unique Opportunity to Understand Hawai'i Volcanoes National Park Ecosystems

Prior to the 1959 Kīlauea Iki eruption, park managers feared that aggressive nonnative plants would in time competitively replace most native plants throughout the park. They were spending considerable time and funds in trying to control the alien invaders. Often their methods of control were not very effective in the long run.

Geologists and vegetation scientists believed that the new volcanic habitats, formed in the Devastation Area, were ideal outdoor laboratories for long term studies on how plants enter barren lands, how plant communities are formed and ecosystems recover. Also the new volcanic substrates were considered a testing ground for studying the behavior of native and nonnative species in early ecosystem development. In addition this new information would provide park resources managers and interpreters with better knowledge of how to protect and interpret the park's outstanding natural and cultural resources. Park managers agreed to the proposal of vegetation research. The Devastation Area study began soon after the 1959 Kīlauea Iki eruption ceased, end of January 1960.

Different types of volcanic materials, destroyed vegetation, and damaged plant remains, were recognized for outlining six habitats. These were mapped in 1960 (see map Figure 2). Plants that survived the eruption soon began to produce new shoots and abundant seed. Also new plants began to invade all six habitats soon after the disturbance. A checklist of plants recorded during the study is appended with scientific and common names and their family associations. Native species dominate the list. Websites showing close-up pictures of all recorded plants (except of the algae, lichens and bryophytes) are given at the end of the plant checklist.

Habitats 1 and 2 were totally abiotic new habitats. Both of these and Habitats 3 and 4 are in the montane rain forest environment. Habitats 5 and 6 are in mesic or seasonal montane environment. These suffered much less impact from the 1959 explosion, and nearly all existing vegetation survived.

The montane rain forest environment at this altitude of 3900 feet (1190 m) has a mean annual temperature of 61°F (16.1°C) and a mean annual rainfall of about 130 inches (3250 mm). Each month receives at least 4 inches (100 mm) of rain in average. The annual fluctuation in mean monthly temperature is about ± 6°C, roughly the same as the daily temperature fluctuation. The seasonal montane environment, near Keanakākoʻi Crater at 3700 feet (1129 m) altitude, has about 1000 mm less rainfall per year and a dry season lasting from mid-May to mid-October during which the average monthly rainfall is less than 100 mm.

FIGURE 2 (RIGHT). Map of 1960 volcanic materials and plant remains that formed six new habitats:

(1) A pāhoehoe rock lava lake on Kīlauea Iki crater floor, 140 acres (56 hectares);

(2) A cinder cone named Puʻu Puaʻi, 47.5 ac (19 ha);

(3) A spatter habitat with snags and surviving trees just east of the cinder cone, 15 ac (6 ha);

(4) A cinder-ash blanket with tree tops sticking out as snags, 77.5 ac (31 ha);

(5) A forest of defoliated, debranched and partially debarked trees that survived, 312.5 ac (125 ha);

(6) A thin ash fallout area beginning around Keanakākoʻi crater and extending into the upper Kaʻū Desert, 657.5 ac (263 ha).

A total affected area of 1250 ac (500 ha). The map shows the original transect design and the depth contours in cm.

KĪLAUEA
CALDERA
KĪLAUEA IKI
CRATER
BYRON LEDGE
HALEMAʻUMAʻU
CRATER
KEANAKĀKOʻI
CRATER
LUA MANU
CRATER
KAʻŪ DESERT
Rain Forest–cM(C)
closed *Metrosideros* forest with *Cibotium* tree fern undergrowth
Seasonal Forest–cM(ns)
closed *Metrosideros* forest with native shrub undergrowth
50 cm
12.5 cm
2.5 cm
North
0 500 1000 m
HABITAT 1
Massive Lava with Joint Cracks
HABITAT 2
Cinder Cone
HABITAT 3
Spatter Area with Tree Snags
HABITAT 4
Cinder-ash with Tree Snags
HABITAT 5
Cinder-ash with Surviving Trees
HABITAT 6
Thin Fallout Area
X Weather Stations

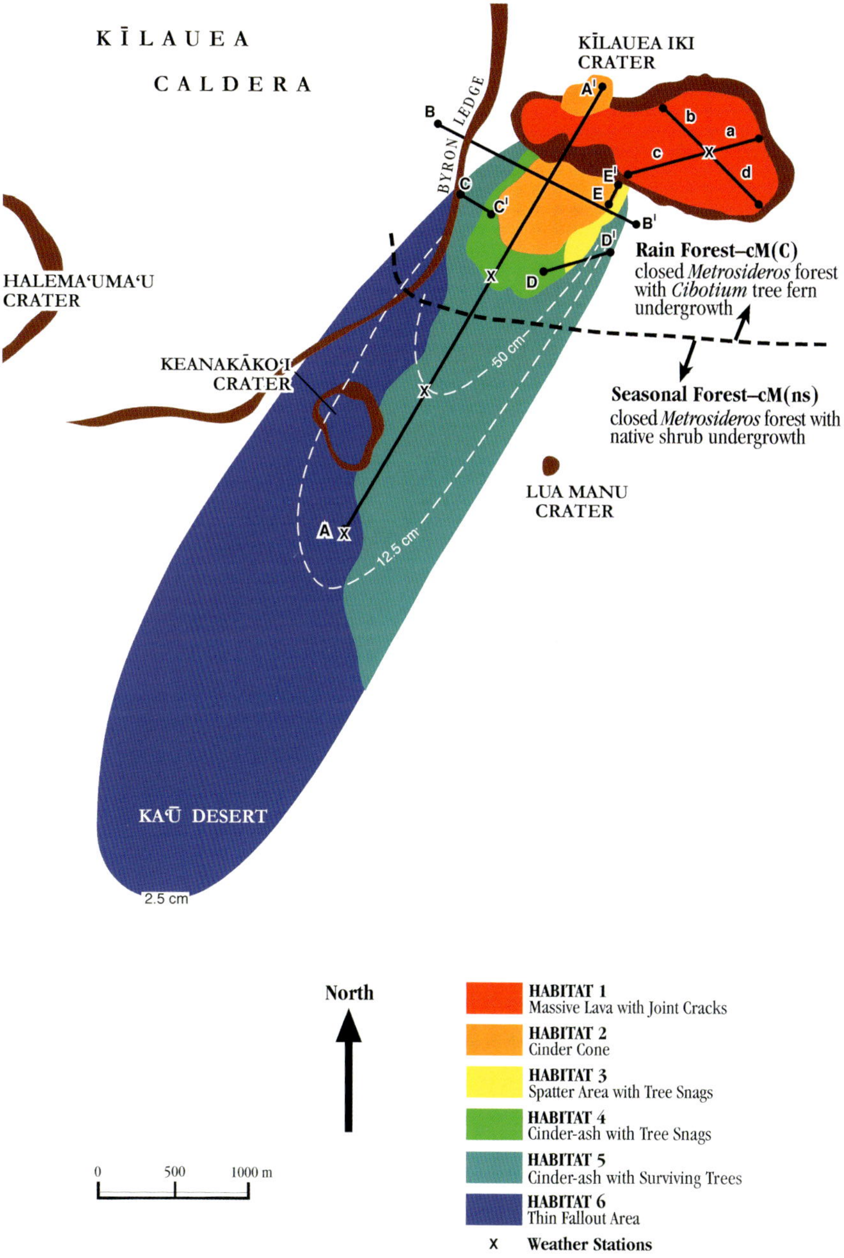

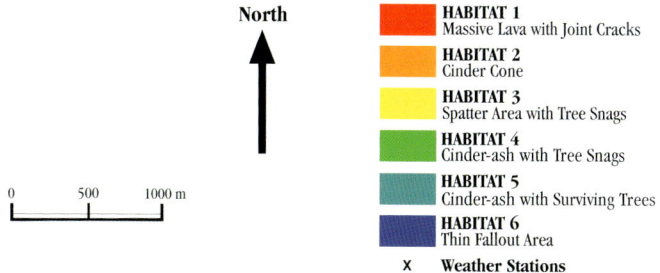

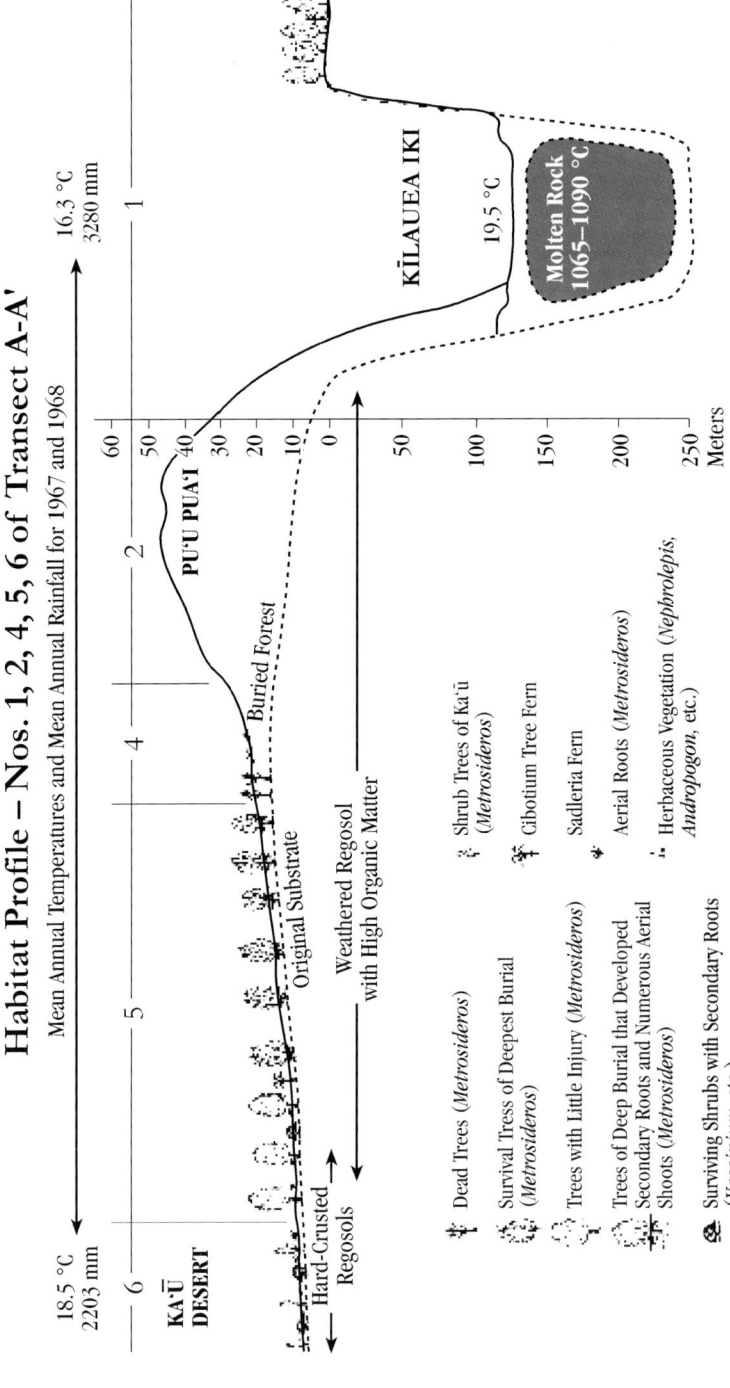

Figure 3. Topographic profile of Devastation Area Habitats extending from upper Ka'ū Desert to Kīlauea Iki along transect A to A' (on Figure 2).

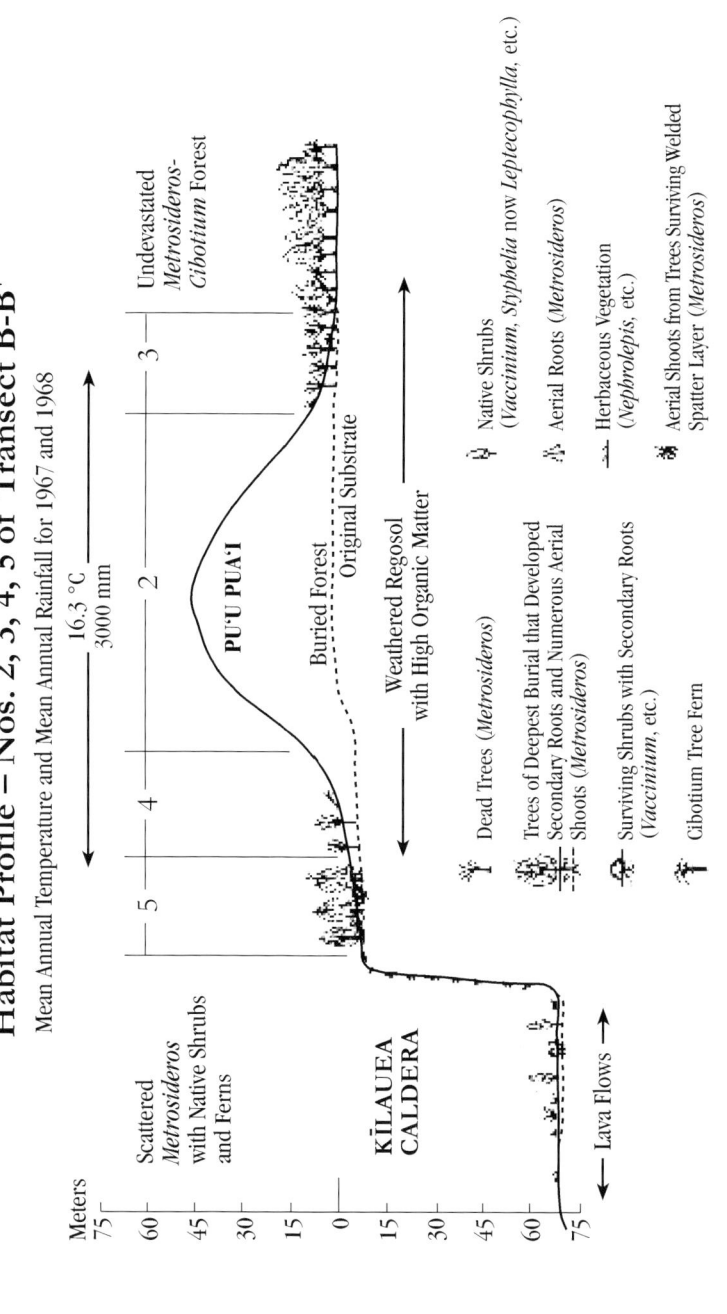

FIGURE 4. Devastation area profile of transect B to B' (Figure 2) extending from Kīlauea caldera across Puʻu Puaʻi to undevastated montane rain forest.

Park scientists began a thorough study of each habitat. Permanent photo stations, belt transects, plots and weather stations were established to record plant recovery and invasion of new plants. Plants were recorded periodically by their percent cover, frequency and density in each habitat. This study will continue for many years into the future, adding new knowledge about native and nonnative plant ecology of Hawaiian ecosystems.

Vegetation Surveys

Each habitat has been studied for plant invasion and vegetation development over a period of 46 years (1960–2006). Quantitative plot and transect surveys were done almost annually up to 1968. The methods were reported in National Park Service Scientific Monograph 5. Subsequently, the plot and transect system was resurveyed at longer intervals (1974, 1981, 1988), and walk-through and photographic surveys were done almost annually for supplementary information.

Invasion of new plant life was rather rapid initially. The sequence of invaders began with algae, mosses, and low growing ferns. Then a lichen, *Stereocaulon vulcani*, appeared in the third and fourth year on the crater floor and cinder cone, which were the two completely abiotic habitats.

Lichens are symbiotic organisms consisting of a fungus and an alga. Together they form a tough life form combination able to withstand prolonged dry periods and great heat by undergoing temporary dormancy. Moreover, *Stereocaulon vulcani* was found to fix Nitrogen from the air, an essential plant nutrient unavailable on abiotic habitats. This lichen was also found to aid in the initial process of weathering of basaltic rock surfaces.

The four plant life forms recorded first are the evolutionary primitive plant life forms, which probably were the pioneer plant life forms long before humans ever came to witness such invasion process. The other four habitats with dead and surviving woody plants showed some early grass and sedge invasion. This could be attributed to the finer, almost soil-like texture of the volcanic ash that came as fall-out material down at the sides and at further distances away from the site of explosion in the ash fall-out area, affecting habitats 3, 5 and 6 in particular. Seedlings of 'ōhi'a lehua (*Metrosideros polymorpha*), the major native forest canopy species in the Park, also arrived early at the crater floor margin within five years after the eruption.

After 1968, when the damaged trees and shrubs recovered in Habitats 3 through 6, the invasion of new plant life slowed down. New ferns and seed plants became established gradually and sometimes more spontaneously, such as the faya tree (*Morella faya*), an alien tree invader, which appeared in the Devastation Area in the early 1970s.

The study has helped park officials understand how native and non-native plants compete with one another during early pri-

PHOTO 9. Scientists recording plant invasion and vegetation development on new volcanic surfaces.

mary succession and recovery after a volcanic eruption. It also provided information of how plant communities are formed and the time it takes for raw volcanic surfaces to become covered by vegetation. This knowledge is leading to a better understanding, interpretation and protection of the park's unique vegetation.

Photo 10. Scientists working on south side of Habitat 2 Puʻu Puaʻi.

Photo 11. The late Dr. Lani Stemmerman assisting in vegetation resurvey on transect BB' in 1981.

Vegetation Development
Habitat 1: The Crater Floor Habitat

PHOTO 12. Forested floor and sides of Kīlauea Iki crater before 1959 eruption (see also aerial Photo 3).

Prior to the eruption, rain forest dominated by ʻōhiʻa lehua trees (*Metrosideros polymorpha*) and hapuʻu tree ferns (*Cibotium* spp.), extended down into the pit crater of Kīlauea Iki. Earlier, lava had poured into this crater in 1868.

Crater floor emplacement. In November 1959, lava flowing from the base of the Kīlauea Iki fountain filled the crater with a 400-feet (122 m) deep lava lake, destroying the montane rain forest on the crater floor as still seen on the photo above. As the lake cooled, the surface solidified, breaking into numerous polygonal slabs with joint cracks and crevices. For the first few years the cracks and joints steamed vigorously. By the 10th year they began to emit less vapors as the lake cooled below the surface. By the 30th year most of the cracks and crevices ceased to steam.

PHOTO 13. Lava flowing into crater burying the former rain forest. Trade wind carried tephra downwind building a cinder cone, thereby burying another rain forest and damaging a seasonally dry forest downwind with a blanket of cinder-ash.

Photo 14. Year-1. After eruption ceased the crater floor became a hot stone desert with no signs of life.

Photo 15. A weather station was established on the crater floor to measure rainfall, fog interception, humidity and temperature.

Photo 16. Two rain gauges, one equipped with Grunow-type fog interceptor.

Photo 17. Plant life invades the new rock.

Vegetation Development: Habitat 1

Photo 18. 'Ōhi'a lehua (*Metrosideros polymorpha*) tree seedling appearing in crevice after four years (1964).

Photo 19. New 'ōhi'a lehua tree seedling 100 feet (30 m) from margin on crater floor in shallow fissure filled with rock splinters. Note naked root exposure shows invasion was unaided by other plants, 1966.

Photo 20. 'Ōhelo 'ai (*Vaccinium reticulatum*) shrub seedling appearing in lava crack next to sword fern, 1964.

Photo 21. Established 'ōhi'a lehua sapling in flower, one of the earliest at the entrance onto the Kīlauea Iki lava floor from Thurston Lava Tube growing on the marginal "bathtub ring," 1981.

Photo 22. The sword ferns became abundant in some parts of the crater floor particularly on the north-east side, 1981.

Invader sequence. The first plant life to invade the new lava rock included a few blue green algae, a lichen, several mosses, and a sword fern, the false 'ōkupukupu (*Nephrolepis multiflora*). The ferns appeared in the crevices already in year two (1962). The lava lake surface was a hot, hostile environment with acid steam arising from numerous polygonal cracks. Some thermophilic microorganisms such as actinomycetes (fungi) were able to inhabit the steaming vents.

Four years after the eruption, native 'ōhi'a lehua (*Metrosideros polymorpha*) tree seedlings and 'ōhelo 'ai (*Vaccinium reticulatum*) shrub seedlings invaded next to the already established ferns in some cooling joints and cracks, especially at the outer rim of the lava rock lake. Blue green algae were the first to appear on the surface of the lava rock. After four years, the lichen *Stereocaulon vulcani* also appeared on the rock surface while mosses and ferns preferred the cracks between the lava slabs. These early plant life forms therefore occupied distinctly different microhabitats on the crater floor. Over the 45 years of observing plant invasion on the crater floor, all additional plant life, in particular the vascular seed plants, assembled only in the lava cracks, not on the surfaces of the slabs.

Soon after 'ōhi'a lehua tree seedlings and 'ōhelo 'ai shrub seedlings became established, other native woody plants such as kūpaoa (*Dubautia scabra*), pilo (*Coprosma* sp.), kopa (*Hedyotis centhranthoides*), māmake (*Pipturus albidus*), and later 'ōhelo kau la'au (*Vaccinium calycinum*) and 'ilihia (*Cyrtandra platyphylla*) appeared in the most sheltered places at the north end of the crater floor.

Densification in cracks. Over time dense colonies of sword ferns formed in cooling joint cracks across the crater floor. These often combined with scattered 'ōhelo 'ai shrub seedlings and scattered 'ōhi'a lehua tree seedlings. Native pioneer ferns such as 'ae (*Polypodium pellucidum*) and 'ama'u (*Sadleria cyatheoides*) also appeared in the lava fissures.

PHOTO 23. Geologists drill into surface of lava lake 1981.

Habitat constraints. Geologists periodically drilled into the lava lake surface to determine how fast its molten interior was cooling. In 1988 (29 years after eruption) drillers found no more melt. Thus, the lake's interior had become solid rock.

The lake surface continues to form new cracks and crevices. Minor earthquakes continue to occur in the interior of the lava lake, and deep within the summit of Kīlauea volcano. These earth movements have reopened some steaming cracks that had previously ceased to steam. Where steaming returned already established plant aggregations disappeared.

PHOTO 24. Some juvenile 'ōhi'a lehua in flower near the center on lava lake, pointed to by Julia Williams. 'Ama'u fern (*Sadleria cyatheoides*) with dried up lower fronds in center of photo, 1998.

For 38 years the lava lake surface remains a harsh pioneer habitat. But the native 'ōhi'a lehua trees and associated native woody plants have steadily progressed towards its center. In addition, the nonnative broomsedge grass (*Andropogon virginicus*) has also invaded and spread in certain crevices where some organic matter has become lodged.

Photo 25. Distant view of Kīlauea Iki crater floor 1998.

Photo 26. 'ōhelo 'ai (*Vaccinium reticulatum*) near center of lava lake in a fissure totally unaided by other plants.

Photo 27. Small juvenile 'ōhi'a lehua trees begin to become obvious on certain parts of the crater floor, 1988.

Dispersal question. What brought 'ōhelo 'ai here? Berries are generally not distributed by wind as are the spores of ferns and the small seeds of 'ōhi'a lehua. However, when berries are dried up, wind may swipe them and their seeds across a smooth surface. We did a seed-trap study in the 1960s, which unfortunately was vandalized. Thus, seed dispersal remains an open question. It is possible that forest birds, such as the endemic 'oma'o (*Myadestes obscurus*) or the alien Japanese white-eye (*Zosterops japonicus*) are involved but so far no such observations have been reported. The role of the introduced Kalij pheasant (*Lophura leucomelana*), whose population recently increased explosively in the Devastation Area and elsewhere in the Park, is currently under investigation.

After 40 years plant life becomes visible from a distance on the crater floor. However, the barren lava surface dominates the view. Plant life has become denser on the outer margin of the lake surface. The observed invasion pattern on Kīlauea Iki crater floor has turned out to be different from that observed earlier on the nearby Keanakāko'i crater floor (see Habitat 6).

Photo 28. Pioneer communities form in various places around the broader margin of the lava lake. Steaming areas are still present and mostly avoided by plants 38 years after the eruption.

Photo 29. Kīlauea crater in 2006, duplicating view of year 1 on Photo 14, page 23. Note the ʻōhiʻa lehua tree seedlings are now well established on the "bathtub ring" of the broken pāhoehoe lava rock.

Vegetation Development
Habitat 2: The Cinder Cone Habitat

PHOTO 30. The cinder cone Puʻu Puaʻi.

Cinder-ash fell from the high lava-fountain around the southwestern rim of Kīlauea Iki crater, building a 152-feet (46 m) high cinder cone, Puʻu Puaʻi—"Hill of High Fountain." At the same time, another rain forest segment was completely buried, leaving a totally abiotic habitat. Puʻu Puaʻi steamed for several years. It became a distinctive landmark by its orange-whitish minerals deposited over its summit. Deep fissures formed in various places across the surface of the cinder cone as the ash hill settled.

Habitat constraints. Habitat 2 is a harsh environment with high daily surface temperature ranges, similar in that respect to Habitat 1. Although the porous soil receives large amounts of rain water, it cannot hold moisture long enough in its surface for most plants to become established and survive. The substrate is moist only when it rains. The hardiest and most adaptable pioneer species can take a foothold on this porous substrate. Cinder are frothy pyroclastic fragments of irregular shape heaped loosely into a pile of any size.

Development of summit community. Blue green algae, mosses and some ferns were sparsely scattered over the cinder top after one year. Increasing numbers of these pioneer plant life forms were joined by several other native species. In the fourth year, the lichen, *Stereocaulon vulcani* was found on the summit. Concurrently, native woody seed plants appeared in the fissures notably a few seedlings of the ʻōhiʻa lehua tree.

By 1969, four native woody plants, ʻōhiʻa lehua, ʻōhelo ʻai, pūkiawe (*Leptecophylla tameiameiae*), and kūpaoa (*Dubautia scabra*) were colonizing surfaces and fissures in the summit area of the cone. Three native ferns soon appeared, ʻae (*Polypodium pellucidum*), ʻamaʻu (*Sadleria cyatheoides*), and uluhe (*Dicranopteris*

Photo 31. Pioneer growth in a summit joint crack with ʻōhiʻa lehua and ʻōhelo ʻai becoming prominent around 1970.

Photo 32. Close-up of a summit fissure on Puʻu Puaʻi. ʻŌhiʻa lehua tree seedling rising from a fissure; kūpaoa shrub seedling attached to its right on rough cinder; a false ʻōkupukupu fern in foreground.

Photo 33. Dense mat of native lichens on Puʻu Puaʻi summit being invaded with native ʻōhelo berry and ʻae fern, 1988.

linearis), the latter restricted to the deeper fissures. Two nonnative ferns, the sword fern, the false ʻōkupukupu (*Nephrolepis multiflora*) and the goldback fern (*Pityrogramma austroamericana*), became more abundant. Seedlings of the nonnative huelo ʻīlio bush (*Buddleia asiatica*) and broomsedge grass made a temporary appearance in the developing summit community.

Twenty years after the eruption, two lichens, *Stereocaulon vulcani* and *Cladonia skottsbergii*, became more prominent, *Stereocaulon* on the rock surface and *Cladonia* on organic plant fragments that were deposited there by wind. The native fern ʻae (*Polypodium pellucidum*) increased in abundance.

Nēnē, a formerly lost link in community formation. After extinction in the wild, nēnē geese were re-introduced into the Park through a captive breeding program in the mid-1960s. By the late ʼ70s to early ʼ80s, nēnē geese were residing in the Devastation Area. Nēnē have been seen landing on the summit of Puʻu Puaʻi and foraging there. They also were commonly observed in the snag habitat adjoining the cinder cone where they were seen to forage on ʻōhelo ʻai. The red and yellow berries of this native shrub are one of their favored food sources. It is likely that nēnē are responsible for distributing ʻōhelo ʻai bushes, which rather suddenly appeared on the slopes of the cinder cone in the early 1980s.

Photo 34. A flock of the native nēnē geese (*Branta sandvicensis*) near the eastern base of Puʻu Puaʻi facing Kīlauea Iki. Here they are feeding on introduced grasses and look for Park visitors. Visitors should not feed them.

Photo 35. ʻŌhelo ʻai bush on slope of cinder cone, a major native food plant of the nēnē.

While the summit of the cinder cone became occupied mostly by native pioneer plants in the first decade, some alien hardy pioneers also became settled there. They include the bamboo orchid, the false 'ōkupukupu sword fern, the broomsedge, (*Andropogon virginicus*), and some small cushions of knotweed (*Persicaria capitata*).

PHOTO 36. The south slope of Pu'u Pua'i in an early stage of invasion, 1974. Some 'ōhelo 'ai bushes became established in front on the cinder surface. A few scattered 'ōhi'a lehua tree seedlings took hold in the cracks on this slope in the distance.

PHOTO 37. A section on the southwest slope of the cinder cone along transect BB', where a pūkiawe shrub individual and an 'ama'u fern became established in 1981. These two species are as yet sparsely represented on the cinder cone.

Development of slope communities. The slopes of the cinder cone remained rather barren for almost two decades after the eruption. The south and southwest slopes became dominated by scattered bushes of ʻōhelo ʻai, starting in the mid-1970s. ʻŌhiʻa lehua seedlings preferred the fissures in the cinder cone.

PHOTO 38. View of the developing ʻōhelo ʻai community on the southwest slope of Habitat 2, the cinder cone. ʻŌhiʻa seedlings are as yet confined to cracks, 1981.

PHOTO 39. Surveying the developing ʻōhelo ʻai shrub community on the cinder cone with Prof. Orlóci and Rhonda Loh along transect BBʹ in 1998.

40 VEGETATION DEVELOPMENT: HABITAT 2

PHOTO 40. 'Ōhi'a seedlings in crack on south slope of cinder cone, 1998.

PHOTO 41. Advanced slope invasion dominated by 'ōhelo 'ai. These shrubs occur in a scattered formation on the south side of the cinder cone in 1998.

PHOTO 42. Typical invasion pattern on the south slope of the cinder cone: scattered 'ōhelo 'ai bushes followed by kūpaoa, the cushion shrub, with white flowers in front on this picture, 1998.

PHOTO 43. A few advanced 'ōhi'a lehua seedlings in sapling stage and many smaller ones seen climbing up on the east slope of the cinder cone 44 years after the explosion. A few scattered broomsedge grasses follow. 'Ōhi'a lehua seedlings show up as silhouettes on the horizon, 2004.

On the east slope upward from the spatter Habitat 3, ʻōhiʻa lehua seedlings invaded slowly, starting on firmly compressed cinder and avoiding as yet the loose rubble below the summit.

Photo 44. Lower east slope of Puʻu Puaʻi showing invasion with many ʻōhiʻa lehua seedlings. A few pukiawe and ʻōhelo ʻai shrubs are present also on this more firmly compressed cinder slope in 2004.

Photo 45. Puʻu Puaʻi in 2006 duplicating view on Photo 30 taken in 1981, page 33. The cinder cone surface is now more densely occupied by native shrubs, mostly by ʻōhelo ʻai and the occasional pūkiawe. ʻŌhiʻa lehua tree seedlings are so far not much in evidence.

Vegetation Development: Habitat 2 43

Photo 46. View from top of Pu'u Pua'i with summit community in foreground, 2004. 'Ōhi'a lehua seedlings in crack in front and on left. 'Ōhelo 'ai shrubs form a scattered assemblage on the right side behind the summit on slope of cinder cone down to the Snag Habitat 4. The latter is identifiable in the background by the 'ōhi'a lehua trees that have in part recovered from snags. The Survival Tree Habitat 5 is the more closed forest at the horizon.

In summary, invasion became most notable first on the summit of Pu'u Pua'i, thereafter on its slopes. Here, first on the southwest side towards Byron Ledge, then on the south side upwards from Habitat 4 and last on the southeast and east side upwards from Habitat 3. The dominating invader was the 'ōhelo 'ai shrub on the cinder surface, whereas the 'ōhi'a lehua seedlings became preferentially established in the cracks of the Cinder Cone Habitat 2.

Vegetation Development
Habitat 3: The Spatter Habitat

PHOTO 47. The Spatter Habitat after installation of the boardwalk in 1960.

Habitat 3 was a small segment of devastated rain forest at the south rim of Kīlauea Iki crater adjoining the Puʻu Puaʻi cinder cone on its eastern side (see map Figure 2). Here the ash came down in a dense hail and formed a blanket up to three feet (90 cm) deep. The term spatter refers to gas-rich, light weight ejecta that are partially liquid when striking the ground. At the ground, the hot, partially liquid fragments usually become welded into an agglutinated mass. When cooled, spatter clods can often be recognized by shiny surfaces of volcanic glass. In the Spatter Habitat the forest was damaged severely and all undergrowth was buried, leaving trees standing as snags. This new volcanic substrate held rain water longer than any other type of pyroclastic material produced in this eruption. The improved water relations allowed for rapid invasion of both native and non-native pioneer plants.

Initial invasion. Blue green algae and sword ferns appeared throughout Habitat 3 in 1961. These were soon followed by exotic grasses such as dallis (*Paspalum dilatatum*), yellow fox tail (*Setaria parviflora*), and rat tail (*Sporobulus africanus*). Blackberry (*Rubus argutus*) shrub seedlings soon joined the thimbleberry bushes (*Rubus rosifolius*). The latter *Rubus* species invaded rapidly through the spatter habitat from east to west towards the boardwalk (now the paved Devastation Trail).

PHOTO 48. Another view of Habitat 3 towards the less disturbed rain forest, 1961.

Some park officials were concerned that the *Rubus* bushes would rapidly spread over the new volcanic surfaces of the Devastation Area. Control measures were taken by spraying with herbicides. The park scientists suggested that removing of the aliens would alter the succession processes. The park superintendent at that time agreed with the scientists and determined that the long-term study of the Devastation Area was to continue for observing successional changes without human interference. Over the next few years, grasses and thimbleberry bushes became replaced in the course of succession by newly invading native and nonnative herbaceous and woody plants.

Photo 49. Two years later in 1962 showing an invading front of the alien thimbleberry (*Rubus rosifolius*) advancing towards the boardwalk in front of photo.

Another spontaneous alien shrub invader, soon causing much concern, was the huelo 'īlio or dogtail, also known as butterfly bush (*Buddleia asiatica*). Many individuals advanced rapidly across the boardwalk in the Spatter Habitat 3 towards the base of the Cinder Cone Habitat 2. But they stopped here and became established only next to the standing snags of the 'ōhi'a lehua.

Photo 50. Four years later, nonnative huelo 'īlio (dogtail, *Buddleia asiatica*) bushes had invaded and grown up quickly at some standing 'ōhi'a lehua snags next to the boardwalk on the side towards the cinder cone.

Favorable microhabitats. The next three photos show favored microhabitats for early pioneer plants in this spatter habitat. These microhabitats are the bases of tree snags and tree molds left from fallen snags. Here both water and nutrient relations were improved relative to the general spatter surface. The standing tree snags intercepted wind driven rain. Some leaf material accumulated at their base and in the tree molds.

Vegetation Development: Habitat 3

Photo 51. A nonnative pāmakane (*Ageratina riparia*) growing at the base of a snag, 1962.

Photo 52. The sword fern (*Nephrolepis multiflora*) together with 'ōhi'a lehua at the base of a snag in deep spatter.

Photo 53. 'Ama'u (*Sadleria cyatheoides*) fern sporeling and mosses in tree mold.

Vegetation Development: Habitat 3

A transient invader. The following three photos demonstrate the role of ʻōhiʻa lehua snags in supporting invasion on the new spatter surface of the alien shrub huelo ʻīlio. It had become apparent that the standing snags functioned as an additional water source intercepting horizontal rain. As soon as one of the snags fell down the affected invader was deprived of the additional water source and began to die. Thus, the huelo ʻīlio bush was not establishing itself without remnants of the previous forest. In other words it turned out not to be an entirely independent pioneer plant.

Photo 54. Tree snags in the spatter area adjoining the cinder cone in 1964.

PHOTO 55. The same snag pair in 1967 with two nonnative huelo 'īlio bushes established at their bases.

PHOTO 56. The same place in 1969. One huelo 'īlio has died where the tree snag had fallen.

The huelo ʻīlio became an aggressive invader in all snag habitats during the first decade. However, after about two decades, its density declined. This decline appeared to be independent of such supporting roles as provided by tree snags. The general decline of the huelo ʻīlio during the 1980s was related to the short lifespan of these shrubs. We found them to be short-lived perennial bushes. They entered a senescing life stage after about twenty years to die soon thereafter. In this way, they have a transient function in providing organic matter to the raw volcanic substrate, which favors invasion of successional plants.

Some snags become alive. Following are two photos that show the ability of half-buried ʻōhiʻa lehua trees, whose tops appeared like snags, to recover vegetatively from their partially buried trunks. Re-sprouting occurred at and a little below the spatter surface and fully foliated trees appeared in the deep spatter at the base of the cinder cone. Initially, they were thought to be sexually reproduced seedlings, but closer inspection revealed their origin as re-sprouts of tree snags formerly believed to be dead.

Some of these recovering snags replaced huelo ʻīlio bushes that had established from seed at their bases.

Photo 57. Vegetatively recovered 'ōhi'a lehua tree snag on the cinder cone side of the board walk in the Spatter Habitat, 1981.

Photo 58. A vegetatively recovered 'ōhi'a lehua snag near the base of the cinder cone in the Spatter Habitat, 1981.

Early succession. A place near the boardwalk similar to Photos 48 and 49 (pages 46 and 47 respectively) shows the recovery of vegetation after 21 and 28 years respectively. The surviving ʻōhiʻa lehua trees are in full foliage from top to base (Photo 59). The great increase in plant biomass is obvious. The thimbleberry bushes have been successionally displaced by other alien and native species. A conspicuous alien is the Japanese anemone (*Anemone hupehensis*) and among the natives are the hāpuʻu tree fern (*Cibotium glaucum*) and shrubs such as ʻōhelo kau lāʻau (*Vaccinium calicynum*), māmake (*Pipturus albidus*) and pūkiawe (*Leptecophylla tameiameiae*). In the foreground of the photo is a row of the alien broomsedge grass (*Andropogon virginicus*), which benefits from the sunlight along the boardwalk. It rarely occurs inside the developing forest.

PHOTO 59. Surviving ʻōhiʻa lehua trees fully refoliated with cylindrical crowns, and some in flower, 1981.

Photo 60. A section of Habitat 3 near the boardwalk, 1988. It shows a patch of Japanese anemone in a recovering 'ōhi'a lehua rain forest.

Photo 61. Pa'iniu (*Astelia menziesiana*), normally growing as a rain forest epiphyte was found as a rare native invader on the spatter surface of Habitat 3 in 1978.

PHOTO 62. Recovery of Spatter Habitat 3 in 2006, matching Photo 47 in 1960, page 45. The boardwalk was unfortunately replaced with a paved walkway. This invited trampling off the defined Devastation Trail causing alien grass invasion combined with additional water moving laterally from pavement and widening of the right-of-way.

PHOTO 63. View of Spatter Habitat 3 closer to Cinder Cone Habitat 2. Note hāpu'u tree fern (*Cibotium glaucum*) in undergrowth left and 'ōhi'a tree seedlings in opening in front. The taller trees are recovered 'ōhi'a lehua snags, some of them still in juvenile form with foliated branches along entire trunk and/or basitonic branching.

Vegetation Development
Habitat 4: The Tree Snag Habitat

Photo 64. Segment of Habitat 4, cinder-ash area with tree snags, 1960

The Tree Snag Habitat differed from the Spatter Habitat by being covered with a loose blanket of cinder-ash. The blanket varied in depth from 40 feet (12 m) at the base of the cinder cone to 1 foot (30 cm) at the boundary to the survival tree Habitat 5. The cinder particles are actually a very light form of ash, generally from less than 1 to 2 cm³ in size. They are very light because volcanic gas had formed internal cavities when molten. During settlement of the cinder-ash blanket, sink holes were formed in a few places by collapse into open fissures in the underlying rocks. All trees sticking out of the blanket appeared to have died as they were recognized as dead tree snags in 1960. This area is still in the rain forest environment but grades into the seasonally dry environment near the border of Habitat 5.

PHOTO 65. This view shows Habitat 4 in the foreground and the survival tree Habitat 5 in the adjacent background in 1981.

In 1981, the survival tree Habitat 5 appeared to be almost fully recovered while the snag Habitat 4 was still in the early stages of succession.

The first significant colonizer. This was the native cushion shrub kūpaoa (*Dubautia scabra*) which appeared already in 1963. In 1974 most of these cushion shrubs were at peak maturity and a few years later they began to senesce and thereafter, to die. These early invaders thereby demonstrated that they had only a short life cycle of about 15–20 years as individual plants in this habitat. The first colonizers behaved as cohorts (members of one generation). Thereafter, reproduction of kūpaoa was less explosive and individuals regenerated at different times.

Photo 66. A cohort colony of kūpaoa (*Dubautia scabra*), a native, dense cushion-forming shrub had invaded the western area of the snag habitat. The kūpaoa cohort colony at peak maturity growing independently among ʻōhiʻa lehua snags in 1974.

PHOTO 67. A young invading huelo ʻīlio bush protrudes from a vigorous kūpaoa mat which begins to die in its center, 1977.

PHOTO 68. A sinkhole in the cinder-ash blanket occupied by an ʻōhiʻa lehua snag, a few sword ferns, and a kūpaoa mat that survived in the collapsed sinkhole, 1988. Sinkholes like this, formed in several places of the snag habitat, showing that the substrate was still settling, thereby remaining unstable over the first 20 years.

Kūpaoa mats facilitating new invaders. The dying cushion shrubs changed into favorable microhabitats for successional species to follow. They formed light gray mats of litter and branchlets flattened near the ground that provided wind shelter and nutrients from decomposing organic matter. They offered no competition to the plants that occupied these mats in succession.

PHOTO 69. Invasion of pink colored knotweed (*Persicaria capitata*) in dead kūpaoa mat, 1988. Knotweed became a new obvious alien invader in the early 1980s.

PHOTO 70. Invasion of 'ōhi'a lehua tree seedling in senescing kūpaoa mat, 1988.

Independent and dependent pioneers. 'Ama'u ferns participated as significant native plant invaders in widely scattered formation in early vegetation development of Habitat 4.

PHOTO 71. A lone native 'ama'u fern at the boundary of the snag habitat in the forefront of the cinder cone, 1977.

Photo 72. A nonnative huelo ʻīlio (dog tail) bush has invaded a dead kūpaoa mat as an early successional plant. Next to it is a native ʻamaʻu fern that established on the raw ash, 1977.

Photo 73. A lone pāwale (*Rumex skottsbergii*) in forefront of the Snag Habitat 4 towards the Cinder Cone Habitat 2, 2002.

PHOTO 74. Pāwale engulfing dying huelo 'īlio bush on Transect AA' in Habitat 4, 1981.

The endemic pāwale shrub (*Rumex skottsbergii*) arrived suddenly with a small population of wind-dispersed individuals in the late 1970s. As new individuals became established in Habitat 4, they moved towards and up the cinder cone as independent pioneers. But the larger individuals soon began to die, indicating that they had a short life cycle of less than 15 years. They never became abundant, but are fast growing transient colonizers, who like the alien huelo 'īlio bush, act as facilitators for other invaders.

Snags become alive. Ōhi'a lehua snag recovery came as a surprise after at least ten years of dormancy. It underscores the concept of "life from the ashes." Such recovered trees contributed organic matter in form of leaf litter which attracted herbaceous species such as the red-colored knotweed (*Persicaria capitata*) a mat-forming introduced plant that now is very evident in Habitat 4. Also the native mat-forming plant kūkaenēnē (*Coprosma ernodeoides*) developed in these microhabitats created by 'ōhi'a lehua as well as seedlings of the faya tree (*Morella faya*).

Photo 75. Surprisingly, a number of snags buried to near their tops under the cinder-ash blanket became refoliated. This demonstrated an enormous capacity of ʻōhiʻa lehua to regain life through vegetative resprouting after 20 years of appearing dead.

Photo 76. Annette Mueller-Dombois pointing at dead huelo ʻilio (*Buddleia asiatica*) shrub and recovering ʻōhiʻa lehua snag in Habitat 4, 1990.

Photo 77. An 'ōhi'a lehua bush that recovered vegetatively from the surviving trunk of a buried tree that appeared to be a dead snag, 1977.

Fallen snags provide favorable microhabitats. Two decades after the eruption, mats of *Cladonia skottsbergii* and *C. oceanides* often expanded their coverage near the fallen snags thereby extending the favorable microhabitat created by fallen snags. These lichens facilitated further vegetation development on the loose cinder-ash blanket.

PHOTO 78. Two lichens participated in the vegetation development on Habitat 4, *Cladonia skottsbergii*, and *Cladonia oceanides*. They began to appear on the fallen snags and increased in abundance with accumulating organic matter around them, 1977.

Photo 79. Year 22. Dense lichens mats on fallen limbs, 1981. Finger points to red-capped *Cladonia oceanides*, also referred to as British soldiers.

Photo 80. View of a favorable microhabitat on the cinder-ash blanket of Habitat 4. A broken down snag left a small depression with decaying organic matter on which a group of invaders became established. The lichen *Cladonia skottsbergii*, the sword fern, the broomsedge, and a dead pāmakani (*Ageratina riparia*) 1977.

Vegetation Development: Habitat 4

Photo 81. Uninformed park visitors thought sinkholes useful for dumping broken snags to clean the "disorderly surface" left by nature. Thereby, they removed favorable microhabitats for plant invasion and also destroyed the *Stereocaulon* lichen cover by trampling over the cinder-ash surface, an unfortunate anthropogenic effect.

Favored nēnē habitat. There are a number of dead and dying huelo ʻīlio bushes on Photo 82 (next page) and a still vigorous individual in the left foreground. Among the fallen snags are dead mats of kūpaoa (*Dubautia scabra*) which are now occupied by *Cladonia* lichens, patches of sword fern, and the small pinkish knotweed (*Persicaria capitata*). At the boardwalk, inside Habitat 4 (extending across middle of Photo 82), are several ʻōhiʻa lehua individuals recovered from half dead snags. They show densely foliated branches from base to top resulting in cylindrical shapes. The cinder cone slope (Habitat 2) appeared still rather barren in 1981. Note the nēnē goose walking among the fallen snags in the foreground. Nēnē are feeding on berries of ʻōhelo ʻai (*Vaccinium reticulatum*) and seem to play a significant role in restocking their own "fruit orchard." The ʻōhelo ʻai bushes became suddenly abundant on the slope beginning in the late 1970s and early 1980s.

Photo 82. View from Snag Habitat 4 across the former wooden boardwalk with photo station (now paved walkway without photo station) towards the cinder cone in 1988. Can you spot the nēnē walking in the foreground?

Photo 83. Another view of vegetation development in the lower part of Habitat 4 closer to the Survival Tree Habitat 5, where the ash blanket has thinned out to less than 2 feet (60 cm) depth, 1988. Here human trampling has compacted the cinder-ash thereby promoting densification of the broomsedge grass.

In the lower part of Habitat 4 (Photo 83), more ʻōhiʻa lehua snags were seen to resprout and become densely refoliated. Their flowers, in turn, attracted the Japanese white-eye (*Zosterops japonicus*) whose droppings brought faya tree (*Morella faya*) seeds. They quickly developed into fast growing trees next to some recovered ʻōhiʻa lehua trees. Raw patches of cinder-ash are still apparent among the denser vegetation which is becoming dominated by the alien broomsedge (*Andropogon virginicus*). The tall growing ferns are native and the low growing cover plants include the native kūkaenēnē (*Coprosma ernodeoides*) and the alien knotweed (*Persicaria capitata*). Among the taller shrubs are the alien blackberry (*Rubus argutus*), some dead and dying huelo ʻīlio bushes, and the native māmake (*Pipturus albidus*). Native and exotic plant life combined in the early vegetation development in this section of Habitat 4, where considerable organic matter had ameliorated the raw cinder-ash in twenty years following the explosion.

Photo 84. Nēnē walking from cinder cone to its favored habitat. Vegetation islands, such as on this photo, often started from single native plant invaders among the fallen ʻōhiʻa lehua snags.

Successional ecotones. Where the Snag Habitat 4 joins the successionally advanced Spatter Habitat 3 one can observe successional ecotones. This term refers to a transition zone, where the earlier sun-loving (heliophytic) invaders become displaced by the later, more shade-tolerant, invaders. The earlier invaders on Photo 85 are the false ōkupukupu fern, the broomsedge, and the knotweed forming the pink patch in the foreground. There is a dying or defoliated rubus bush, and behind that are successional plants that tend to come in later. They include the mat-forming native uluhe fern (*Dicranopteris linearis*) and next to it the tall Japanese anemone. This succession process is promoted by the ameliorating microclimate in Habitat 4 near the spatter Habitat 3. Here the closing tree canopy of ʻōhiʻa lehua is the primary factor in ameliorating the microclimate.

Photo 85. Photo showing a number of key species in the process of early- to mid-succession in Habitat 4, 1988. In the foreground center is the false ʻōkupukupu fern (*Nephrolepis multiflora*). In front at right is a patch of the pinkish knotweed and at both sides the broomsedge grass.

Another successional ecotone is shown on the following photo at the southeast corner, where the Snag Habitat 4 joins the Survival Tree Habitat 5. Here a rather negative development has occurred. The loose cinder-ash blanket has become compacted by trampling. This in turn has promoted a dense stand of broomsedge grass, a fire hazard. Moreover, the Survival Tree Habitat became densified by the alien faya tree (*Morella faya*). Seeds of the faya tree were brought in by birds, notably the Japanese white-eye (*Zosterops japonicus*). The birds were attracted by the flowers of the ʻōhiʻa lehua tree.

PHOTO 86. A close-up view from the open Snag Habitat into the Survival Tree Habitat.

In the foreground is a grass cover formed by broomsedge (*Andropogon virginicus*). Behind the grass, a small ʻōhiʻa lehua tree is engulfed by fast growing faya (*Morella faya*) trees. This indicates two threats for the survival of ʻōhiʻa lehua. The broomsedge is a pyrophyte, meaning it attracts and survives grass fires. The alien faya tree is a fast growing nitrogen fixer which can overshadow and thereby kill the ʻōhiʻa lehua tree. Because of its aggressiveness under these conditions, it became a target of control by Park Management in Habitat 5.

PHOTO 87. Cinder-ash area with tree snags in 2006, showing a similar segment in Habitat 4 as on Photo 64, 1960, page 59. In front some broken down snags; several were removed unfortunately and thrown into sink holes by uninformed visitors. The cushion shrubs in front are mostly the endemic kūpaoa (*Dubautia scabra*). The bushy trees behind, forming a spatially open structure, are recovered snags of the endemic ʻōhiʻa lehua (*Metrosideros polymorpha*). Note their complete refoliation from top to bottom. This is characteristic of physically injured trees (a similar response as to hedge cutting), and a setback to juvenility, causing increased carbon dioxide absorption due to increased leaf area.

Vegetation Development: Habitat 4

Photo 88. Another segment of Habitat 4 with recovered and unrecovered snags. The pink patch surrounding the broomsedge bunchgrass (*Andropogon virginicus*) in front is the now widely distributed alien knotweed (*Persicaria capitata*). The shrub on the left below the leaning snag is the endemic hanupaoa (*Dubautia ciliolata*), which sometimes occurs among the more common kūpaoa (*Dubautia scabra*).

Vegetation Development
Habitat 5: The Survival Tree Habitat

PHOTO **89.** The Survival Tree Habitat after deposition of the ash blanket gives the appearance of dead and dying trees in year 1, 1960.

The Survival Tree Habitat 5 was situated between Habitats 4 and 6 along the ash fallout gradient which also happens to be associated with a decreasing rainfall gradient. The ash fallout blanket varied in depth from 10 feet (3 m) to less than one foot (30 cm) and consisted of increasingly finer and lighter cinder-ash along the AA' gradient (see map Figure 2 and profile Figure 3). The trees of this seasonally dry or mesic forest lost most of their foliated branches from the abrasive fallout of the cinder-ash.

The dim-looking aftermath. When the explosion ceased in December 1959, most of the branches and undergrowth vegetation appeared to be buried. In the following weeks many additional branches fell to the surface during strong winds. The climate of Habitat 5 is not only characterized by decreasing annual rainfall, but also by increasing dry periods during the summer. After the trees were barren, solar radiation was often rather intense on the surface thereby causing strong desiccation except when low clouds drifted through the skeleton forest.

The tree trunks looked gray from adhering fine ash. The deadly appearance did not last very long. Already at the end of the first year many trees produced adventitious buds along the grayish tree stems and remaining branches.

Life returned quickly. Not only did the trees become heavily refoliated, they also produced abundant flowers at the end of the second year. Park visitors marveled at the color of the recovering 'ōhi'a lehua and some referred to them as "red-hot pokers."

Photo 90. This view shows the rapid recovery by refoliation along the trunks from the top to the bottom of the damaged 'ōhi'a lehua trees in year 2, 1962.

Photo 91. Typical abundance of blossoms on a surviving 'ōhi'a lehua tree that recovered quickly in Habitat 5 in year 2.

The flowering trees produced an abundance of seeds which were wind-dispersed over a wide range throughout the Devastation Area. The seeds of 'ōhi'a lehua are very small and light and easily carried by wind over long distances. However, to germinate and become seedlings, they require favorable micro-habitats with moisture enduring in the surface. As long as the ground is under the influence of strong desiccation, seedlings cannot develop.

A fertilizing effect? The cross-section of the injured 'ōhi'a lehua tree displays a surprising boost in diameter growth six years after deposition of the cinder-ash blanket (Photo 92). Was this a fertilization effect from volcanic ash? Most likely not, as the soil nutrients were still tied up in crystallized form. Small amounts of volcanic glass could have released some potassium early. But the boost in diameter growth can be said to have been an indirect fertilization effect. The boost was likely related to the sudden increase in foliage all over the tree trunks, which resulted from the development of adventitious buds promoted by physical injury.

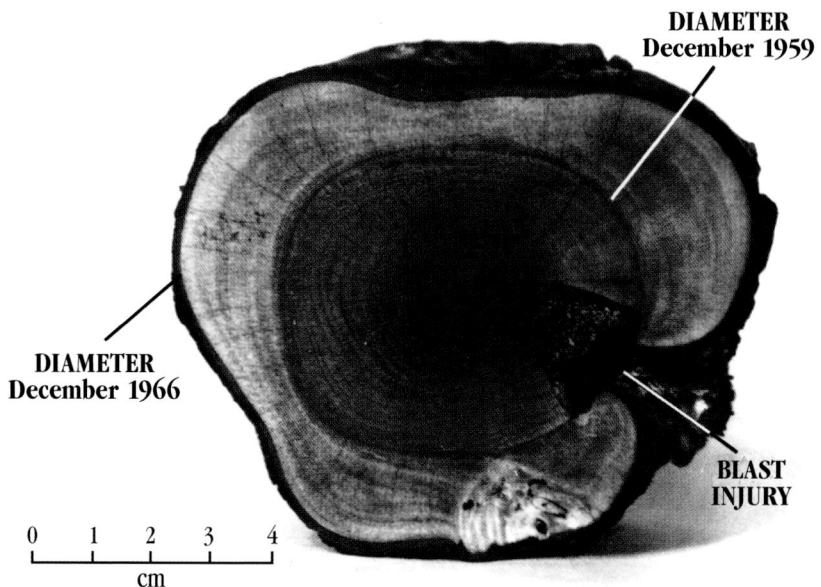

Photo 92. Cross section of 'ōhi'a lehua stem from surviving stand (Photo 90, page 80), Habitat 5.

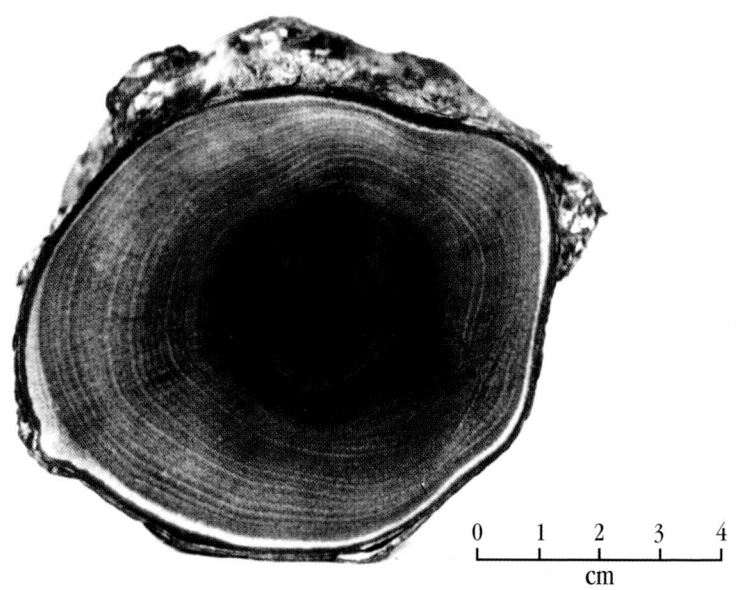

Photo 93. Cross section of 'ōhi'a lehua stem of uninjured tree in forest adjacent to Habitat 5. It also grew faster after acquiring more stem foliage following crown defoliation from ash fallout.

The trees became enshrouded with thickly foliated branches and assumed a cylindrical shape. Such shape is typical for juvenile 'ōhi'a lehua trees. In this case the damaged trees were vigorous mature trees prior to the ash fall-out. Their increased leaf area after physical damage is most likely responsible for the sudden fertilizing effect through additional CO_2 absorption from the air.

An adaptive response? Aerial roots appeared on a few surviving trees that were buried under the deeper 6–9 feet (2–3m) ash blanket in Habitat 5. Some scientists hypothesized that these thick broom-like aerial roots served as respiration organs in place

Photo 94. The photo shows some surviving 'ōhi'a lehua trees with reddish dense, broom-like aerial roots in year 7, 1967.

PHOTO 95. A few aerial roots made contact with the soil and others dried up already in year 7.

of buried root systems, at least temporarily. Unfortunately, this hypothesis was not tested at that time.

By year 20 (1981) most of the aerial roots, including those that had reached the ground, had fallen from the trees or had dried up. Obviously, their function was only temporary. Most likely adventitious roots had developed in the meantime under ground from the buried trunk nearer the surface of the ash overlay. Such adventitious rooting of ʻōhiʻa lehua trunks has been observed elsewhere at blown-out ash dunes in the Kaʻū Desert.

Herbaceous invaders. Several species of native and nonnative herbaceous plants, which grew from bulbs or rhizomes below ground level, survived the ash burial where the blanket was a foot (30 cm) or less thick. The nonnative species included some garden escapees, such as the five ornamental species below. Among these were the Kāhili ginger (*Hedychium gardnerianum*), the montbretia (*Crocosmia x crocosmiiflora*), the Chinese ground orchid (*Phaius tankarvilleae*), the Philippine ground orchid (*Spathoglottis plicata*), and the bamboo orchid (*Arundina graminifolia*). A native plant in this category was the tall 'uki sedge (*Machaerina angustifolia*).

The Kāhili ginger has become one of the most invasive undergrowth plants in some sections of the Park's rain forest. It was introduced as an ornamental from the Himalayas and first collected in the Park about 1940.

PHOTO 96. Nonnative kāhili ginger or 'awapuhi kāhili (*Hedychium gardnerianum*).

PHOTO 97. Nonnative montbretia (*Crocosmia x crocosmiiflora*).

Montbretia originated through hybridization of two species in the genus *Crocosmia,* both of them native to eastern South Africa. It spreads only vegetatively but is widely distributed in transported soil. It is often seen along roadsides and the edges of trails.

The Chinese ground orchid (*Phaius tankervilleae*) is one of the world's oldest cultivated orchids. It is native to southern China. After being introduced to Hawai'i it escaped cultivation. It is now found growing in the wild in many open places, usually not inside forests.

Photo 98. Nonnative Chinese ground orchid.

Photo 100. Nonnative bamboo orchid.

Photo 99. Nonnative Philippine ground orchid with broomsedge grass behind.

The Philippine ground orchid (*Spathoglottis plicata*) is a native of Malaysia. It was introduced to Hawai'i as an ornamental, but soon escaped cultivation. It invades open moist areas and can persist in the shade of forest canopies.

The bamboo orchid (*Arundina graminifolia*) was introduced to Hawai'i from the East Indies and escaped cultivation. It is found throughout the seasonal dry forest. This species as well as the broomsedge grass can be displaced by an advancing mat of the native uluhe, the false staghorn fern, in rain forest succession.

Shrub invaders. A few, nonnative blackberry (*Rubus argutus*) began to invade in an open area at the fringe of the Survival Tree Habitat in year 3. By year 7 they had become well established. Park officials were concerned that this aggressive nonnative shrub would take over the Devastation Area, "driving the native plants out."

PHOTO 101. Blackberry bushes in an open area at the Survival Tree Habitat, 1963.

PHOTO 102. Rich crops of blackberry brambles attracted berry pickers, year 3.

Park visitors and Volcano Village residents often visited the Devastation Area, picking the large juicy berries to make pies and jelly. Also nēnē geese were observed to feed on these exotic berries.

By year 14, many of the earlier blackberry bushes began to senesce and were invaded by the native kukaenēnē (*Coprosma ernoideoides*). The small shiny black berries of kukaenēnē are an original food item of the nēnē geese.

Following recovery of the forest canopy after 38 years, the undergrowth also had recovered well with many of the native shrubs becoming more dense, particularly pūkiawe (*Leptecophylla tameiameiae*) and ʻaʻaliʻi (*Dodonaea viscosa*). In many places the alien broomsedge (*Andropogon virginicus*) died from lack of light and was replaced by the kūkaenēnē (*Coprosma ernodeoides*), as depicted in the following two views into Habitat 5.

Photo 103. The native creeping shrub kūkaenēnē had spread vigorously around some of the surviving trees by year 14.

Photo 104. Vigorous undergrowth of kukaenēnē developed around the bases of some fully recovered 'ōhi'a lehua trees in Habitat 5, 1998.

Tree invaders. Seedlings of the alien tree species faya (*Morella faya*) rapidly developed into trees because of their nitrogen-fixing capacity. This was a serious problem of alien tree invasion. Systematic control measures were taken by Park Resource Management, and Habitat 5 was declared a "Special Ecological Area."

In Habitat 4, the Snag Habitat, faya trees were left growing for longer observation to determine how they compete with the recovering vegetation in this volcanically more severely disturbed area. So far, no 'ōhi'a lehua tree has been displaced by the faya tree in this habitat. It was hypothesized that the faya tree would attract more alien weeds because of its nitrogen-fixing ability. This hypothesis was tested by a special scientific study in Habitat 4 and not found to be supported.

Some koa trees were planted along the Crater Rim and Chain-of-Craters Road in the seasonal dry forest prior to the 1959 eruption. A few juvenile koa trees invaded after the eruption. Like the dominant 'ōhi'a lehua, the planted koa trees survived and juvenile koa developed from cable-like (stoloniferous) roots away from parental trees vegetatively as clonal trees. These specialized cable-like roots are flexible pencil thin stems that run hard beneath the ground surface, often several meters away from the parent tree. Actually, the N-fixer koa would be the logical successional tree in the more open-structured seasonal ʻōhiʻa lehua forest. But its natural dispersal is limited unlike that of the new N-fixer, the faya tree.

PHOTO 105. Young koa (*Acacia koa*) tree, a native invader following the disturbance in the Survival Tree Habitat 5, 1998.

Photo 106. Faya tree seedlings, as the one shown in this photo, appeared rather suddenly in many places of the Survival Tree Habitat 5 in the early 1970s.

PHOTO 107. Advanced growth of faya pushing through a crown of 'ōhi'a lehua, 1978.

Photo 108. Two saplings of nitrogen fixing trees, the native koa and the alien faya in Survival Tree Habitat 5, 1988.

Photo 109. A segment of the Survival Tree Habitat 5 after full recovery of its canopy in year 38.

Vegetation Development: Habitat 5

Photo 110. Recovered seasonal forest in Survival Tree Habitat 5 at edge of 1974 lava flow. Note the columnar juvenile structure has given way to the more mature crown structure by thinning of lateral branches on most of the stems of the ōhi'a lehua trees, 2002.

Photo 111. Segment of Survival Tree Habitat 5 in 2006 in similar location on transect AA' as Photos 89 (1960, page 79) and 90 (1962, page 80). Most of the 'ōhi'a lehua trees have shed their lateral branches, an indication that they are now fully recovered. A robust endemic undergrowth ground cover of kūkaenēnē (*Coprosma ernodeoides*), as here shown, has become conspicuous in many places. The black berries of kūkaenēnē are also a traditional food source of the nēnē (*Branta sandsvicensis*).

Vegetation Development
Habitat 6:
The Thin Fallout Area

Photo 112. Weather station in Habitat 6 which was serviced from 1960 through 1969.

Habitat 6 was the least affected area from the 1959 explosion. The ash blanket varied from 1 foot (30 cm) down to 1 inch (2.5 cm) depth along the AA' transect in this area which extends from the Survival Tree Habitat 5 southward into the upper Ka'ū Desert (see map Figure 2).

The desert concept. Because of its scant vegetation and mostly barren surface, this habitat was designated "desert." However, it is not a "climatic desert." Instead, it is a "substrate desert" due to its recent volcanic age and substrate constraints. Under the loose fallout blanket is a partially cemented substrate layer that originated from a phreatic explosion of the Halema'uma'u crater in 1924. In addition, the site is at the drier end of the rainfall gradient across the Devastation Area from north to south (see Figure 3).

A new opportunity. The damage during the 1959 cinder-ash fallout in this area was limited to some shearing-off and burning of shoots of the woody plants and to the burying of their lower stem areas. Even low growing shrubs were only partially buried. Only a few widely scattered, drought resistant native plants and plant assemblages were growing in cracks and crevices of the hard volcanic surface prior to the arrival of the new shallow cinder-ash blanket. Thus, the loose overlying ash blanket offered a new opportunity for vegetation development in this area.

The area receives little rain fall during the summer months. It is exposed to high solar radiation, and constantly blowing dry winds. The northeast trade winds, which unload most of their moisture over the rain forest, have a desiccating effect on this

PHOTO 113. A section of the upper Ka'ū Desert where plants were buried under one foot of ash. Some shrubs were burned to ground level and most were stripped of foliage.

area. Yet, the climate is sufficiently moist. It could support a seasonally dry forest, were it not for recency of the soil substrate and its additional impedimentation from cementation now just beneath the surface.

PHOTO 114. A closer view of upper Ka'ū desert in 1963, three years after the ash blanket was deposited. Native shrubs have resprouted from buried root and branch systems.

Adventitious rooting. The branches of the individual (in Photo 115, page 100) were sheared off and had burned during the fallout event. The shrub recovered rapidly by developing new adventitious roots from the stem in the new cinder-ash layer and its old root system survived. The branches were refoliated by the third year in 1963. Similar stem-based lateral rooting into the new ash blanket was found in 'ōhi'a lehua shrubs (Photo 116).

Photo 115. A native 'ōhelo 'ai bush that regained vitality by developing adventitious roots from the buried stem.

Photo 116. Close-up of an 'ōhi'a lehua shrub that was buried under almost one foot (30 cm) of ash and survived similarly by developing lateral roots from the lower stem, which grew into the new ash blanket.

Shrub islands. The next three photos show shrub-island communities built around one or two stunted 'ōhi'a lehua trees in the upper Ka'ū Desert. The associated shrubs are native. They include pūkiawe (*Leptecophylla tameiameiae*), 'a'ali'i (*Dodonaea viscosa*), 'ōhelo 'ai (*Vaccinium reticulatum*), and kūkaenēnē (*Coprosma ernoidoides*).

The shrub islands indicate the harsh environment of the underlying hard-pan and the desiccating northeast trade winds.

PHOTO 117. Native shrub islands in the upper Ka'ū Desert. Kīlauea caldera with Mauna Loa in background.

The 'ōhi'a lehua also plays a role in intercepting horizontal rain that comes through this area frequently during the cooler season of the year thereby providing additional moisture to the shrub island community in this desert habitat.

Photo 118. The stunted and stag-headed 'ōhi'a lehua indicates its adaptation to grow in this harsh environment of Habitat 6.

Photo 119. A desert shrub island with a patch of the tall native 'uki sedge pointed to by Garrett Smathers under the dead 'ōhi'a lehua tree, 1998.

Shrub-island associates. The following photos depict native and non-native key species frequently associated with the desert shrub islands. The faya tree is a recent invader, whose invasion was facilitated by the loose cinder-ash blanket.

PHOTO 120. A close-up of the pūkiawe (*Leptecophylla tameiameiae*) shrub showing its needle-like leaves and berries. This is an important native shrub in the desert shrub islands associated with 'ōhi'a lehua.

PHOTO 121. Close-up of a flowering branch of the 'a'ali'i (*Dodonaea viscosa*), another important native shrub in the desert shrub islands associated with 'ōhi'a lehua. These two native shrub species are also common in the undergrowth of the Survival Tree Habitat 5.

PHOTO 122. In the foreground from right to left kūkaenēnē, 'a'ali'i, and 'ōhelo 'ai with broomsedge grass and the taller 'ūlei shrub (*Osteomeles anthyllidifolia*) reaching above, 1998.

Photo 123. An occasional native hanu-paoa (*Dubautia ciliolata*) individual in Habitat 6, 2004.

Photo 124. The alien faya tree invaded several desert shrub islands otherwise dominated by native woody plants in the early 1970s.

The faya tree has become locally dominant in this section of the upper Kaʻū Desert, by taking advantage of the long established native community and the new ash blanket. In year 3 (1963), the alien broomsedge grass (*Andropogon virginicus*) invaded first, and then the faya tree appeared displacing the broomsedge at the fringes of the native shrub-island communities. The latest invader was the knotweed (*Persicaria capitata*) which began to form a new fringe around some shrub island communities.

Shifting ash and desert wadis. Near the southern end of transect AA' the cinder-ash layer was uniformly shallow after the eruption, less than four inches (10 cm) deep. Subsequently, the ash was shifted by water during flash floods into gentle depressions. Forty years of observations have revealed that flash-flooding occurs episodically during the winter and spring seasons in both the upper and lower Kaʻū Desert. These desert wadis were created prior to the 1959 ash fall-out. Eroding flood waters carried loads of fine ash from the 1959 cinder-ash layer mixed with gravel from the former hardpan subsoil and deposited this mixed material alongside the streambeds, which are normally dry.

PHOTO 125. Dry streambed-like features, so-called wadis, created by flash flooding in the Kaʻū desert. Note absence of broomsedge grass, 1967.

The next view shows the newly invaded alien broomsedge grass, but also occurring are two native plants, the tall ʻuki (*Machaerina angustifolia*) sedge and small hanu-paoa (*Dubautia ciliolata*) shrub, and more frequently the nonnative bamboo orchid (*Arundina graminifolia*).

Photo 126. A sparse cover of mostly alien invaders settled along the margins of the dry streambeds, 1998. The broomsedge grass became established in the mid-1970s.

Photo 127. Broomsedge invasion confined to wadi on more compacted gravelly ash soil along AA' transect near end iron stake in 1988.

The Keanakākoʻi crater floor. This crater was in the thin fallout area close to the survival tree habitat. When the crater floor vegetation was analyzed in 1962, very little damage was recorded. The surviving trees of ʻōhiʻa lehua were only 5 to 8 feet (1.5 to 2.5 m) tall and rather bushy in appearance. They established on the lava surface of the 1877 flow. Thus, the trees were around 80–85 years old and still in their juvenile life stage.

According to the uniform distribution pattern of ʻōhiʻa lehua trees, the lava surface must have cooled evenly after the flow was imbedded. This crater is in the montane seasonal environment and therefore no tree ferns established on the Keanakākoʻi crater floor, but at least ten native species were associated with the dominant ʻōhiʻa lehua in form of aggregations.

The tree bases were surrounded with various ferns, including the native ʻae (*Polypodium pellucidum*), moa (*Psilotum nudum*), the native sedge, uki (*Machaerina angustifolia*), and several native woody species including pūkiawe (*Leptecophylla tameiameiae*), aʻaliʻi (*Dodonaea viscosa*), ʻōhelo ʻai (*Vaccinium reticulatum*), and

PHOTO 128. Distant view of vegetation on the floor of Keanakākoʻi crater in 1962.

Photo 129. Vertical view into crater in 1967 showing the uniform distribution of the 'ōhi'a lehua tree cohort established about 90 years ago.

kūpaoa (*Dubautia scabra*). A few solitary pāwale (*Rumex skottsbergii*) were also present. Some alien species such as the goldback fern (*Pityrogramma austroamericana*), huelo 'īlio bush (*Buddleia asiatica*), and one forb, the gosmore daisy (*Hypochoeris radicata*) were associated with the native species aggregations.

Photo 130. Closer view of plant growth on Keanakāko'i crater floor, 1962. In average there were around 400 juvenile 'ōhi'a lehua trees per hectare.

PHOTO 131. A closer view of Keanakākoʻi crater floor vegetation, showing ʻōhiʻa lehua tree islands and the raw cinder-ash blanket forming the matrix in 1962 after the ash fallout in 1959.

Keanakākoʻi versus Kīlauea Iki. The Keanakākoʻi crater floor represents 85–90 years of vegetation development in the montane seasonal environment. While its species assemblages are similar to those on the Kīlauea Iki crater, the pattern of development is quite different. After 46 years, the surface of the Kīlauea Iki crater is still rather barren, particularly in the center towards the south end. This pattern of slow recovery is no doubt related to the way the lava surface has cooled, which was not uniform. According to its location in the rain forest environment, one would have expected a more rapid development in the Kīlauea Iki than in the Keanakākoʻi crater.

The crater floor vegetation prior to the 1959 eruption in the Kīlauea Iki crater was about the same age as that of the Keanakākoʻi crater in 1967. A comparison of the 90-year-old rain forest vegetation of the Kīlauea Iki (Photo 12) with the 90-year-old seasonal rain forest vegetation on the floor of Keanakākoʻi crater clarifies that it is not safe to predict vegetation development on outwardly similar habitats. In this case, the different development is related to the dynamics of the substrate.

On July 19, 1974, an eruption occurred along a line of fissures extending from the southern edge of Kīlauea caldera and across the Keanakākoʻi Crater. Lava poured into the Keanakākoʻi crater floor, destroying the 90-year-old plant cover. This disturbance regime maintains the vegetation in the Kīlauea summit area in a pioneer state whereby native species are usually the first colonizers.

PHOTO 132. Keanakākoʻi crater in 1974 after destruction of the 90-year-old vegetation through a new lava flow. Note abscence of any steaming.

Photo 133. Keanakākoʻi crater in 2002 still showing a dominantly barren lava surface after 28 years of exposure.

A New Lava Flow

In 1974 new lava flowed from a fissure that opened at the east side of the Kīlauea Caldera. The fissure extended through Keanakākoʻi Crater and the lava flowed across the southern part of the Survival Tree Habitat into the small pit crater, Lua Manu. This created a new habitat in the Devastation Area (see map Figure 5, which also shows the adjusted transect system).

A different kind of flow. In contrast to the Kīlauea Iki lava lake, the 1974 lava flow was thin. Instead of massive dense and degassed lava, as in the Kīlauea Iki crater, this lava consisted of "shelly" lava, a type of inflated void-rich lava characterized by very thin and fragile lava plates overlying gas cavities.

PHOTO 134. The 1974 lava flow created a number of lava trees indicating that the flow was rapid and then settled during cooling in some places by about two meters, 1998.

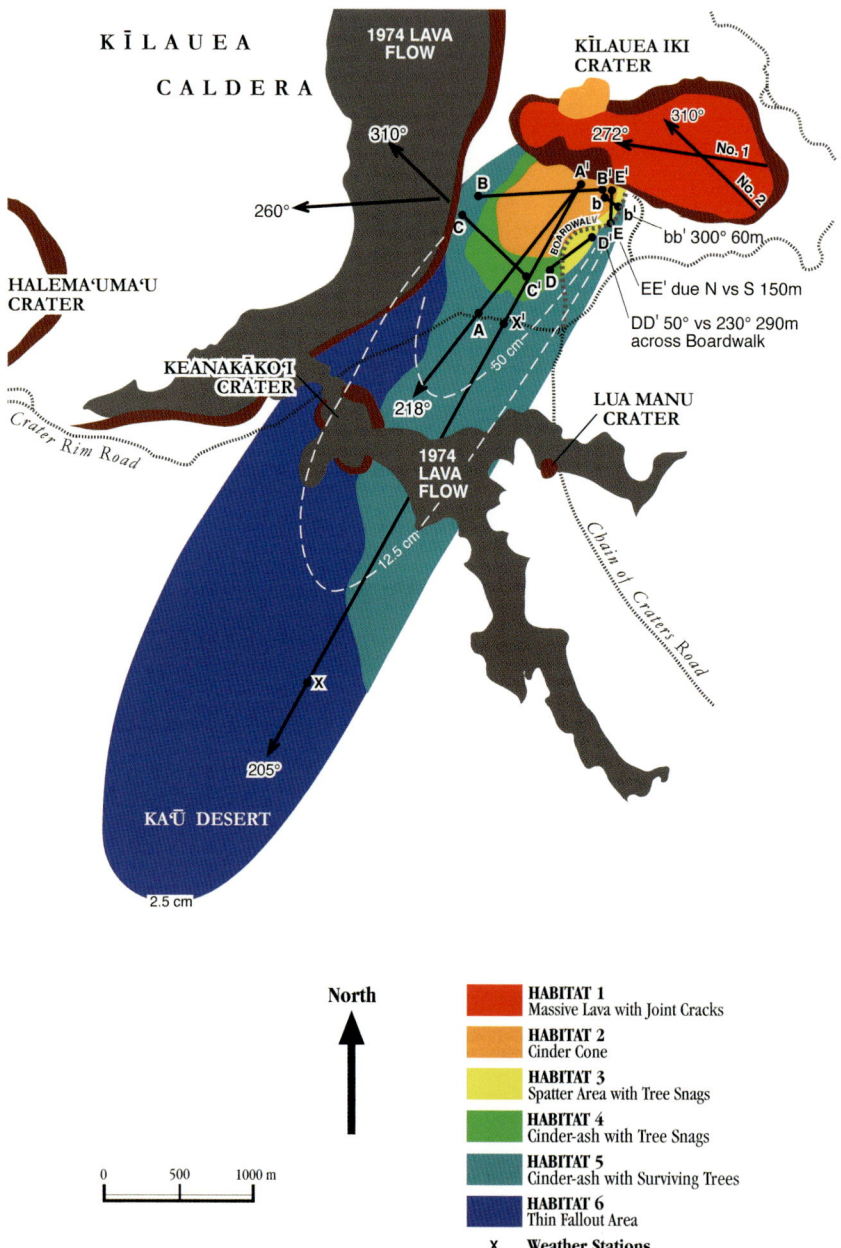

FIGURE 5. Map of Devastation Area with new lava flow of 1974 and transect system adjusted in 1981.

The initial invasion and recovery on the 1974 lava flow was much more rapid than on the Kīlauea lava lake, in spite of the drier climate in the seasonal environment where the lava flow settled. The early invaders on the 1974 lava flow were essentially the same as those on the lava lake. However, the false ʻōkupukupu fern and native woody plants, in particular ʻōhiʻa lehua, kūpaoa, and ʻōhelo ʻai, became more densely and uniformly established across the area. This difference is attributable to the different lava type.

Photo 135. The shallow lava flow did not totally destroy the overrun section of the survival tree forest. Some trees survived and re-emerged through breaks in the shelly lava, others were surrounded leaving small vegetation islands, called kīpuka, as here in background, 1998.

Ōhiʻa lehua survivors. Surprising was the discovery that many early ʻōhiʻa lehua trees came from vegetative recovery pushing from below through the fissures in the shelly pāhoehoe lava. Also, many new seedlings of ʻōhiʻa lehua became established at the same time.

This is a totally different kind of vegetation recovery as compared to the lava flow of the Kīlauea Iki floor.

Photo 136. A vegetatively recovered 'ōhi'a lehua has pushed through the overlaying lava and is here shown in a profused flowering state in 1988.

Photo 137. Garrett Smathers points at a dead endemic pāwale bush (*Rumex skottsbergii*) during the 1998 study. This pioneer shrub is not so common. It has a fast growing rate and a short life cycle, even shorter than the kūpaoa cushion shrub of which there are a number of individuals shown on the above photo.

Summary and Conclusions

Following the volcanic explosion of December 1959, the name Devastation Area was applied to the entire 1250 ac (500 ha) area that was either covered by lava or with a blanket of ash. This extended from Kīlauea Iki into the upper Ka'ū Desert over a distance of 4.1 miles (6.5 km). The vegetation of this area was either destroyed or severely damaged. Today, the effects of devastation are restricted to the Crater Floor Habitat 1, the Cinder Cone Habitat 2, the Spatter Habitat 3, and the Tree Snag Habitat 4. The Survival Tree Habitat 5, and the upper Ka'ū Desert Habitat 6 have recovered from the 1959 devastation.

Habitat 3, which is at the entrance of the Devastation Trail from the southern rim of Kīlauea 'Iki and the southeast side of the Pu'u Pua'i Cinder cone, is now in an early seral state of rain forest succession. So far, the undergrowth of the recovered 'ōhi'a lehua trees in this habitat, is still dominated in many places by the Japanese anemone (*Anemone hupeensis*) but the native hapu'u tree ferns (*Cibotium glaucum*) and the uluhe fern (*Dicranopteris linearis*) begin to re-assume their dominance. The hāpu'u is more evident in shaded positions under the forest canopy and the uluhe fern in a few areas with canopy gaps. Native shrubs in this successionally advanced rain forest are the 'ōhelo kau la'au (*Vaccinium calycinum*), the māmake (*Pipturus albidus*), and the pūkiawe (*Leptecophylla tameiameiae*). Three alien species that are threatening to change the outcome of succession to a less native plant assemblage in the understory are the introduced kahili ginger (*Hedychium gardnerianum*), the Himalaya raspberry (*Rubus ellipticus*), and the palmgrass (*Setaria palmifolia*). They require attention by Resource Management.

We have learned that vegetation development is remarkably fast after volcanic eruptions, wherever ʻōhiʻa lehua trees are not totally buried under a blanket of ash or under a shelly type of lava flow. If the ʻōhiʻa lehua trees were in a vigorous life stage, they have a remarkable ability to recover and facilitate the re-emergence of other plant life.

Vegetation development was much slower in Habitats 1, 2, and 4. Habitats 1 and 2 presented totally abiotic surfaces in 1960. Today both support only a sparse assortment of native and introduced plants. Their barren volcanic surfaces are still the dominating landscape feature.

On the Kīlauea ʻIki lava floor (Habitat 1), plant invasion progressed from the margin to the center much more slowly than expected. The explanation for this became readily evident. It was the slow cooling of the underlying lava, which to this day has prevented plant invasion around its still steaming area in the southern center near the vent. The floor has cooled from the margin inward and is invaded with an assortment of plants. Among the more obvious plants in lava cracks are the false ʻōkupukupu fern (*Nephrolepis multiflora*), shrub seedlings of the native ʻōhelo ʻai (*Vaccinium reticulatum*), and tree seedlings of ʻōhiʻa lehua (*Metrosideros polymorpha*). In the lower parts of many cracks one also finds certain mosses such as *Rhacomitrium lanuginosum*, and *Campylopus* spp. These were the first invaders followed closely by the lichen *Stereocaulon vulcani* on the lava surfaces. This assembly of plants in the fissures progressed over time towards the center of the lava lake. ʻŌhiʻa lehua and ʻōhelo ʻai often grew independently of the mosses and ferns but also commonly in association with them. The native woody seed plants ʻōhiʻa lehua and ʻōhelo ʻai were able to establish in favorable micro-habitats independently of other plants. No alien woody seed plant was found that had such pioneering ability.

As the lava surface began to weather, finely textured rock flakes and dust particles accumulated in cracks and fissures around the margin of the lava lake and the introduced broomsedge grass (*Andropogon virginicus*) became well established, now running in colonies along many cracks. Native fern invaders include the ʻae (*Polypodium pellucidum*) and an occasional ʻamaʻu (*Sadleria cyatheoides*). Initial invasion was rapid and dense along much of the crater rim where the lava surface had settled by about 2 meters after cooling. Here, so called "bathtub rings" developed, exposing numerous splits and fissures, the ideal micro-habitats for early hardy invaders. The most dense and best developed pioneer community grew on the lava bathtub ring at the entrance trail that slopes down into the crater from Thurston Lava Tube. Here, at least eight native seed plants became established, including some individuals of kopa (*Hedyotis centranthoides*), an uncommon but hardy endemic perennial forb. The establishment of ʻōhelo ʻai away from the crater margin presents a puzzle as the berries are not adapted to wind dispersal. However, dried berries may be swept up by wind. It is also quite possible that human dispersal has played a role here, since it is customary to throw ʻōhelo ʻai berries into the crater to please the Volcano Goddess Pele.

The Puʻu Puaʻi Cinder Cone Habitat 2 received the first plant invaders on its summit. This was unexpected. The summit of the cinder cone appeared to be an unlikely site for invaders to become settled first because of its strongly wind-exposed position. However, cracks developed after settling on the cinder cone providing sheltered positions for pioneer plant life. Next to the mosses (*Rhacomitrium* and *Campylopus* species), the hardy false ʻōkupukupu fern appeared early and almost simultaneously the native ʻae fern as well as the ʻōhelo ʻai and ʻōhiʻa lehua. Surprisingly, the native uluhe fern was seen in some wind sheltered cracks after twenty years and kūpaoa (*Dubautia scabra*) appeared on the side of cracks. After senescing, *Cladonia* lichens settled on the dead branches of kūpaoa. Also the native sedge ʻuki (*Machaerina angustifolia*) and the introduced bamboo orchid (*Arundina angustifolia*) appeared after twenty years. Subsequently, the broomsedge grass (*Andropogon virginicus*) arrived and lately also the knotweed (*Persicaria capitata*). The latter had formed two small 4 to 8 inch (10 to 20 cm) large patches by 1998.

Surprisingly, the cinder cone slopes remained almost completely barren for the first decade. We expected 'ōhi'a lehua seedlings to establish first, but their appearance lagged much behind that of 'ōhelo 'ai. This species appeared more or less suddenly in the mid-1970s and settled apparently quite independently on the raw cinder surface, while 'ōhi'a lehua seedlings tended to prefer the cracks that had opened after settling of the cinder cone. Even in these micro-habitats these native tree seedlings did not yet become abundant in the 45 years after the disturbance. The rather sudden establishment of the 'ōhelo 'ai on the slope of the cinder cone coincided with the appearance of the nēnē geese observed in the adjacent Snag Habitat 4. It seemed that the nēnē were responsible for planting their own fruit orchard in a nearly uniform pattern with increasingly shorter distances among individual plants uphill on the slope of the cinder cone. The 'ōhelo 'ai berries are not expected to fly uphill (in contrast to the light, wind-dispersed seeds of the 'ōhi'a lehua and the kūpaoa). Wind dispersal of 'ōhelo 'ai berries on to the south slope of the Cinder Cone is rather unlikely since the prevailing wind comes from the opposite direction. This presumed nēnē/'ōhelo 'ai relationship deserves further study. Reintroduction of this native bird may have brought back a function of biodiversity, a "missing link," that was lacking in the process of vegetation development, after demise of the nēnē in the wild.

Finally, the Snag Habitat 4 which was created leeward of the Cinder Cone Habitat 2 from deposition of the thick and dense cinder-ash blanket provided the most interesting vegetation dynamics during early succession. Here, the prevalent primary invader was the native cushion shrub, kūpaoa (*Dubautia scabra*). This shrub appeared suddenly in the mid-1970s at the southwest side from the border of the Survival Tree Habitat 5, taking hold on the raw cinder-ash surface. Many kūpaoa cushion shrub individuals became established among the snags and in the forefront towards the cinder cone. These initial invaders formed a "cohort" assemblage, a group of colonizers of the same generation. Many individuals grew into circular mats of about one square meter area. After 15 to 20 years they started to senesce and die, often from their centers outward. These senescing and dying kūpaoa mats provided improved micro-habitats for a second wave of invaders. They included those found in the cracks of the Kīlauea Iki lava floor and others, for example the Indian paintbrush (*Castilleja arvensis*), a foreign hemi-parasite and other transient weeds such as flora's paintbrush (*Emilia sonchifolia*), purple cudweed (*Gamochaeta purpurea*), annual fire weed (*Erechtites valerianifolia*), lani wela (*Conyza canadensis*), pua kaia (*Cirsium vulgare*), and *Epilobium* spp. New kūpaoa shrub seedlings developed from the first generation cohort in a more continuous fashion covering some more ground on the barren cinder-ash surface but not any more explosively. The behavior of kūpaoa was similar to that of the huelo ʻīlio (*Buddleia asiatica*), first explosively invading and then gradually declining. Another important native shrub pāwale (*Rumex skottsbergii* and also *Rumex giganteus*) was suddenly invading the Snag Habitat 4 and moving with a few individuals upslope on the Cinder Cone Habitat 2 in the mid-1980s. It persisted into the mid-1990s and gradually declined in similar fashion as the above two species. In all three cases their decline did not seem to be related to a change in climate but to an inert behavior in their population dynamics. They are short-lived perennials that senesce after a few years and also show a decline in the sexual reproductive capacity in their offspring at the same habitat. Another surprising fact was the re-emergence of foliated branches from some of the deeply buried ʻōhiʻa lehua snags believed to be killed. This, among other features, demonstrated the behavior encapsulated in the initial title of this book "Life from the Ashes." Among these vegetatively recovering snags, appeared increasing numbers of ʻōhiʻa lehua seedlings in Habitat 4, typically in favorable micro-habitats.

In summary, the following interactions have been observed:

1. In many cases native pioneer plants facilitated the invasion of introduced plants. The kūpaoa shrub facilitating the invasion of native and introduced plants, and the 'ōhi'a lehua supporting the invasion of the faya tree, can be considered as outstanding examples.

2. The opposite, whereby introduced plants gave support to native invaders was also observed. Here the outstanding example is the false 'ōkupukupu, which usually was the first invader after mosses to attract the hardy native seed plants. Another example is the low-growing kūkaenēnē shrub which penetrated and overran shrubs of senescing blackberry bushes and invaded patches of knotweed that had formed as low cover plants around the crown perimeter of recovered 'ōhi'a lehua trees.

3. The invasion of so many alien forbs leads to the conclusion that early vegetation development on new volcanic surfaces proceeds now more rapidly than prior to the participation of species that became naturalized after human introduction. These herbaceous weed species have a complementary rather than competitive effect on the native invaders. The native plants tolerate the foreign species and the foreign weeds tolerate the native colonizers as long as the vegetation cover is not yet closed. At the same time, the raw habitat becomes more rapidly enriched with organic matter and nutrients. Most of the weeds will disappear when a forest canopy is re-formed. This process will definitely occur in the next 50 to 100 years on Habitats 1, 2, and 4, provided that the area is not again disturbed by a volcanic explosion prior to forest establishment.

4. Introduced species of concern among grasses are the pyrophytes. They include in particular the broomsedge (*Andropogon virginicus*), the bush beardgrass (*Schizachy-*

rium condensatum), and the molasses grass (*Melinis minutiflora*). The latter two grasses are in the area but not yet abundant. They cannot develop in shaded environments. The broomsedge however, is in the forefront and wherever it becomes dense it presents a fire hazard. This happened in the Snag Habitat 4 in the lower southeast area, where human trampling aided in compacting the cinder-ash. Such compaction facilitated densification of the alien broomsedge grass. Pyrophytes can change the disturbance regime to a fire cycle which then may prevent the re-establishment of forest.

5. An introduced species of concern among the 31 forbs listed in the checklist is the kāhili ginger (*Hedychium gardnerianum*) which will persist under rain forest canopy with a tendency to out-compete most other species in the undergrowth.

6. The study has shown that native and introduced species are not two ecologically distinct plant categories that can always be looked at as competitors. In the process of invasion and succession they mostly complement each other in vegetation development. The outcome is ultimately decided by the tall growing species that make up the forest canopy and those that remain or invade the undergrowth, because of their shade tolerance.

7. The ʻōhiʻa lehua tree will definitely remain a dominant species in the area as long as new volcanic substrates are formed and no tree is introduced that is capable to compete in early primary succession on lava flows. The faya tree (*Morella faya*) came close to that, but so far, it has rarely been observed to enter lava flows in Hawaiʻi and its invasion depends on already established plants that attract its alien seed disperser, primarily the Japanese White Eye (*Zosterops japonicus*).

Checklist of Plants Recorded During the Study in the Devastation Area

I = **Indigenous:** species that arrived naturally without human help

E = **Endemic:** species that originated from indigenous ones. Both I & E are native

X = **Introduced:** nonnative species; also referred to as naturalized, exotic, foreign, or alien species

Bluegreen Algae

I *Anacystis montana* now *Gloeocapsa atrata* (Microcystaceae)
I *Coccochloris stagnina* now *Aphanothece stagnina* (Synechococcaceae)
I *Hapalosiphon laminosus* now *Fischerella laminosa* (Fischerellaceae)
I *Scytonema myochrous* now *Conferra myochrous* (Scytonemataceae)
I *Stichococcus subtilis* now *Klebsormidium subtile* (Klebsormidiaceae)
I *Stigonema panniforme* now *Scytonema panniforme* (Scytomataceae)

Lichens

I *Cladonia skottsbergii* (Cladoniaceae)
I *Cladonia oceanides* (Cladoniaceae) British soldiers
I *Stereocaulon vulcani* (Stereocaulaceae)
I *Usnea* sp. (epiphytic lichen) bearded lichen

Mosses, Hornwort and Liverwort

I *Anthoceros* sp. (Anthocerotaceae) hornwort
I *Bazzania* sp. (Lepidoziaceae) liverwort
I *Brachymenium exile* (Bryaceae)
I *Bryum apiculum* now *Bryum mildeanum* (Bryaceae)
I *Bryum argentum* var. *lanatum* (Bryaceae)
I *Bryum crassicostatum* now *Bryum hawaiicum* (Bryaceae)
E *Campylopus densifolius* now *Campylopus hawaiicus* var. *densifolius* (Dicranaceae)
I *Campylopus exasperatus* (Dicranaceae)
E *Campylopus purpureoflavescens* now *Campylopus hawaiicus* var. *hawaiicus* (Dicranaceae)
I *Campylopus umbellatus* (Dicranaceae)
I *Ceratodon purpureus* (Dicranaceae)
I *Ctenidium decurrens* now *Ectoprothecium decurrens* (Hypnaceae)
I *Dicranum spreirophyllum* (Dicranaceae)
I *Ectoprothecium sandwichense* (Hypnaceae)
X *Hypnum plumaeforme* (Hypnaceae)
I *Isopterygium albescens* (Hypnaceae)
I *Leucobryum gracile* (Leucobryaceae)
I *Macromitrium owaihense* now *Micromitrium microstomum* (Orthotrichaceae)
I *Macromitrium piliferum* (Orthotrichaceae)
I *Rhacomitrium lanuginosum* var. *pruinosum* (Grimmiaceae) now var. *lanuginosum*
I *Racopilum cuspidigerum* (Racopilaceae)
I *Rhizogonium spiniforme* now *Pyrrhobryum spiniforme* (Rhizogoniceae)
I *Taxitelium mundulus* (Hypnaceae)
I *Thuidium plicatum* now *Thuidium cymbifolium* (Thuidiaceae)
I *Trichostelium hamatum* now *Radulina hamata* (Sematophyllaceae)
I *Sematophyllum caespitosum* now *Sematophyllum hawaiiense*
I *Weissia viridula* now *Weissia controversa* (Pottiaceae)

Ferns and Fern Allies

I *Asplenium adiantum–nigrum* (Aspleniaceae) ʻiwaʻiwa
E *Cibotium glaucum* (Dicksoniacae) hāpuʻu
X *Cristella dentata* (Thelypteridaceae) paiʻiʻihā, formerly *Cyclosorus dentatus*
I *Dicranopteris linearis* (Gleicheniaceae) uluhe, false staghorn fern
I *Lepisorus thunbergianus* (Polypodiaceae) pākahakaha, formerly *Pleopeltis thunbergiana*
I *Lycopodiella cernua* (Lycopodiaceae) wawae ʻiole, formerly *Lycopodium cernuum*
E *Nephrolepis exaltata* subsp. *hawaiiensis* (Nephrolepidaceae) true ʻōkupukupu
X *Nephrolepis multiflora* (Nephrolepidaceae) false ʻōkupukupu, formerly misidentified as *N. exaltata* subsp. *hawaiiensis*, true ʻōkupukupu
X *Pityrogramma austroamericana* (Pteridaceae), goldback fern
X *Pityrogramma calomelanos* (Pteridaceae), silverback fern
E *Polypodium pellucidum* (Polypodiaceae) ʻae
I *Psilotum nudum* (Psilotaceae) moa
E *Sadleria cyatheoides* (Blechnaceae) ʻamaʻu
I *Sphenomeris chinensis* (Lindsacaceae) palaʻā, paʻapalaʻā, formerly *S. chusana*
X *Phymatosorus grossus* (Polypodiaceae) false lauaʻe, false maile-scented fern, formerly *P. scolopendria*
E *Pteridium aquilinus* var. *decompositum* (Polypodiaceae)

Grasses, Sedges, Rushes

I *Agrostis avenacea* (Poaceae) heʻupueo
E *Agrostis sandwicensis* (Poaceae) pili-hale
X *Andropogon virginicus* (Poaceae) broomsedge grass
X *Anthoxanthum odoratum* (Poaceae) sweet vernal grass
X *Axonopus fissifolius* (Poaceae) narrow leaved carpet grass

- X *Bulbostylis capillaris* (Cyperaceae), a small annual sedge
- E *Carex wahuensis* (Cyperaceae), a native sedge
- X *Cynodon dactylon* (Poaceae) mānienie haole, Bermuda grass
- I *Cyperus polystachyos* (Cyperaceae), a slender annual sedge
- X *Cyperus rotundus* (Cyperaceae) kiliʻoʻopu, nut grass
- E *Deschampsia nubigena* (Poaceae) hair grass
- X *Digitaria fuscescens* creeping kūkaepuaʻa
- X *Erharta stipoides* (Poaceae) meadow rice grass, formerly *Microlaena stipoides*
- I *Gahnia gahniiformis* (Cyperaceae), a perennial sedge
- X *Holcus lanatus* (Poaceae) common velvet grass
- X *Kyllinga brevifolia* (Cyperaceae) kiliʻoʻopu, formerly *Cyperus brevifolius*
- E *Luzula hawaiiensis* (Juncaceae), a native rush
- I *Machaerina angustifolia* (Cyperaceae) ʻuki
- X *Melinis minutiflora* (Poaceae) molasses grass
- X *Melinis repens* (Poaceae) Natal redtop grass, formerly *Rhychelytrum repens*
- X *Paspalum dilatatum* (Poaceae) dallis grass
- X *Paspalum urvillei* (Poaceae) vasey grass
- X *Pennisetum clandestinum* (Poaceae) kikuyu grass
- X *Sacciolepis indica* (Poaceae) Glenwood grass
- X *Schizachyrium condensatum* (Poaceae) bush blue grass, formerly *Andropogon glomeratus*
- X *Setaria parviflora* (Poaceae) yellow foxtail, formerly *Setaria gracilis*
- X *Setaria palmifolia* (Poaceae) palm grass
- X *Stenotaphrum secundatum* (Poaceae) ʻakiʻaki haole, buffalo grass
- X *Vulpia myuros* (Poaceae) rat tail fescue

Forbs

- X *Ageratina riparia* (Asteraceae) pāmakani haole, formerly *Eupatorium riparium*

- X *Anemone hupehensis* var. *japonica* (Ranunculaceae) Japanese anemone
- X *Arundina graminifolia* (Orchidaceae) bamboo orchid, formerly *A. bambusifolia*
- E *Astelia menziesiana* (Liliaceae) kaluaha, pua ʻakuhinia, paʻiniu
- X *Bidens pilosa* (Asteraceae) kī nehe
- X *Castilleja arvensis* (Scrophulariaceae) Indian paintbrush
- X *Cirsium vulgare* (Asteraceae) pua kaia, bull thistle
- X *Crocosmia x crocosmiifolia* (Iridaceae) montbretia
- X *Commelina diffusa* (Commelinaceae) honohono
- X *Conyza canadensis* (Asteraceae) lani wela, horseweed
- X *Cuphea carthagenensis* (Lythraceae) tarweed
- I *Dianella sandwicensis* (Liliaceae) ʻukiʻuki
- X *Emilia sonchifolia* (Asteraceae) flora's paint brush
- X *Epilobium ciliatum* (Onagraceae), unbranched perennial herb, formerly *E. adenocaulon*
- X *Epilobium billardierianum* (Onagraceae), many-branched perennial herb with rose-purple to white flowers, formerly *E. cinereum,* willow herb
- X *Erechtites valerianifolia* (Asteraceae) annual fire weed
- X *Fragaria vesca* var. *alba* (Rosaceae) European woodland white strawberry
- X *Gamochaeta purpurea* (Asteraceae) purple cutweed, formerly *Gnaphalium purpureum*
- X *Geranium homeanum* (Geraniaceae) cranesbill
- E *Hedyotis centranthoides* (Rubiaceae) kopa
- X *Hedychium coronarium* (Zingiberaceae) ʻawapuhi keʻo keʻo, white ginger
- X *Hedychium gardnerianum* (Zingiberaceae) kāhili ginger
- X *Hypericum parvulum* (Clusiaceae) St. John's wort
- X *Hypochoeris radicata* (Asteraceae) gosmore daisy
- I *Lythrum maritimum* (Lythraceae) pūkāmole
- X *Mentha spicata* (Lamiaceae) kepemineka, spearmint

X *Oxalis corniculata* (Oxalidaceae) 'ihi'ai, yellow wood sorrel, possibly a Polynesian introduction
X *Oxalis corymbosa* (Oxalidaceae) 'ihi pehu, pink wood sorrel
X *Phaius tankarvilleae* (Orchidaceae) Chinese ground orchid
X *Physalis peruviana* (Solanaceae) pohā berry, short-lived perennial sub-shrub
X *Persicaria capitata* (Polygonaceae) knotweed, formerly *Polygonum capitatum*
X *Plantago major* (Plantaginaceae) laukahi, broad-leaved plantain
I *Solanum americanum* (Solanaceae) pōpolo, glossy nightshade
X *Sonchus oleraceus* (Asteraceae) pualele
X *Spathoglottis plicata* (Orchidaceae) Philippine ground orchid
X *Veronica plebeia* (Scrophulariaceae) trailing speedwell

Shrubs

X *Buddleia asiatica* (Buddleiaceae) huelo 'īlio, dog tail or butterfly bush
E *Coprosma ernodeoides* (Rubiaceae) kūkaenēnē
E *Cyrtandra platyphylla* (Gesneraceae) 'ilihia
I *Dodonaea viscosa* (Sapindaceae) 'a'ali'i
E *Dubautia ciliolata* (Asteraceae) hanu-paoa
E *Dubautia scabra* (Asteraceae) kūpaoa
X *Pluchea carolinensis* (Asteraceae) sour bush, formerly *Pluchea symphytifolia*
E *Pipturus albidus* (Urticaceae) māmake
X *Pyracantha crenatoserrata* (Rosaceae) firethorn bush
X *Rubus argutus* (Rosaceae) 'ōhelo 'ele'ele, prickly Florida blackberry, formerly *R. penetrans*
X *Rubus ellipticus* (Rosaceae) yellow Himalayan raspberry
X *Rubus rosifolius* (Rosaceae) 'ōla'a, thimbleberry
E *Rumex giganteus* (Polygonaceae) pāwale
E *Rumex skottsbergii* (Polygonaceae) pāwale
I *Leptecophylla tameiameiae* (Epacridaceae) pūkiawe, formerly *Styphelia tameiameiae*

- E *Vaccinium calycinum* (Ericaceae) ʻōhelo kau lāʻau
- E *Vaccinium reticulatum* (Ericaceae) ʻōhelo ʻai

Trees

- E *Acacia koa* (Fabaceae) koa
- E *Coprosma ochracea* (Rubiaceae) pilo
- E *Coprosma rhynchocarpa* (Rubiaceae) pilo
- E *Metrosideros polymorpha* var. *polymorpha* and var. *glaberrima* (Myrtaceae) ʻōhiʻa lehua
- I *Ilex anomala* (Aquifoliaceae) kāwaʻu
- X *Persea americana* (Lauraceae) avocado
- X *Psidium cattleianum* (Myrtaceae) waiawī, strawberry guava
- X *Morella faya* (Myrtaceae) faya tree, formerly *Myrica faya*
- E *Myrsine lessertiana* (Myrsinaceae), kōlea lau nui
- E *Wikstroemia sandwicensis* (Thymeliaceae) ʻākia

Websites for pictures of:

1. Hawaiian native plants
   Visit the University of Hawaiʻi at Mānoa Botany Department at http://www.botany.hawaii.edu/faculty/carr/natives.htm.

2. Native and nonnative plants
   Visit the Hawaiian Ecosystems at Risk (HEAR) project at http://www.hear.org/plants.

3. Bluegreen algae
   Visit AlgaeBase version 4.0, National University of Ireland, Galway at http://www.algaebase.org.

Photo Credits

Cover photo of fire fountain. From J.P. Eaton, USGS Hawai'i Volcanoes Observatory.

Photo 2. Kīlauea Iki fountain 1959. From J.P. Eaton, USGS Hawai'i Volcanoes Observatory.

Photo 3. Aerial view of Kīlauea summit. From USGS Hawai'i Observatory.

Photo 4. Start of 1959 eruption. From NPS Hawai'i Volcanoes National Park.

Photo 12. Kīlauea Iki Crater prior to 1959. From NPS Hawai'i Volcanoes National Park.

Photo 13. Erupting fountain 1959. From NPS Hawai'i Volcanoes National Park.

All other photos from authors.

Figure Credits

Figure 1. Location map. From Author's NPS Monograph No. 5, 1974.

Figure 2. Devastation habitat map. From Author's NPS Scientific Monograph No. 5, 1974.

Figure 3. Topographic profile along transect AA'. From NPS Scientific Monograph No. 5, 1974.

Figure 4. Topographic profile along transect BB'. From NPS Scientific Monograph No. 5, 1974.

Figure 5. Devastation habitat map with transect system adjusted in 1981 and 1974 lava flow transferred from Stearns (1985).

Bibliography

D'Antonio, C. M., R. F. Hughes, M. Mack, D. Hitchcock, and P. M. Vitousek. 1992. Biological invasion by exotic grasses, the grass/fires cycle, and global change. *Annual Review of Ecology and Systematics* 23: 63–87.

D'Antonio, C. M., R. F. Hughes, M. Mack, D. Hitchcock, and P. M. Vitousek. 1998. The response of native species to removal of invasive exotic grasses in a seasonally dry Hawaiian woodland. *Journal of Vegetation Science* 9: 699–712.

Doty, M. S. and D. Mueller-Dombois. 1966. *Atlas for Bioecology Studies in Hawaii Volcanoes National Park*. Botanical Science Paper No. 2. University of Hawaii, Honolulu, Hawaii. Republished 1970 as Hawaii Agric. Exp. Stat. Misc. Publication 89. 507 pp.

Black, J. M., J. Prop, J. M. Hunter, F. Woog, A. P. Marshall, and J. M. Bowler. 1994. Foraging behaviors and energetics of the Hawaiian Goose (*Branta sandvicensis*). *Wildfowl* 45: 65–109.

Haugen, R. T. 1960. Report of establishment and monitoring photo stations in 1959 Kīlauea Iki pumice fallout area from April–September 1960. Unpublished typescript with handwritten notes. Hawaii Volcanoes National Park Library. 26 pp.

Lamoureux, C. H. 1976. *Trailside Plants of Hawaii's National Parks*. Hawaii Natural History Association, Hawaii Volcanoes National Park. 80 pp.

Matson, P. A. 1990. Plant-soil interactions in primary succession at Hawai'i Volcanoes National Park. *Oecologia* 85: 241–246.

Macdonald, G. A., A. T. Abbott, and F. L. Peterson. 1986. *Volcanoes in the Sea: The Geology of Hawaii. (2nd ed.). University of Hawai'i Press. Honolulu. 517 pp.

Macdonald, G. A. and D. H. Hubbard. 2001. *Volcanoes of the National Parks in Hawai'i.* Updated by C. Haliker and D. Swanson. Hawai'i Natural History Association. Hawai'i Volcanoes National Park. 64 pp.

Motooka, P., L. Castro, D. Nelson, G. Nagai, and L. Ching. 2003. *Weeds of Hawai'i's Pastures and Natural Areas.* College of Trop. Agric. and Human Resources. University of Hawai'i at Mānoa. Honolulu, Hawai'i. 184 pp.

Mueller-Dombois, D. 1983. Population death in Hawaiian plant communities: a causal theory and its successional significance. *Tuexenia* 3: 117–130.

Mueller-Dombois, D. 1986. Impoverishment in Pacific Island forests. Pages 199–210 *in: The Earth in Transition.* G. M. Woodwell (ed.). Cambridge University Press. 530 pp.

Mueller-Dombois, D. 1992. A natural dieback theory, cohort senescence as an alternative to the decline disease theory. Pages 26–36 in *Forest Decline Concepts.* P. D. Manion and D. Lachance (ed.). APS Press. St. Paul, Minnesota. 249 pp.

Mueller-Dombois, D. 2000. Rain forest establishment and succession in the Hawaiian Islands. *Landscape and Urban Planning* 51: 149–157.

Mueller-Dombois, D., K. W. Bridges, and H. L. Carson (eds.). 1981. *Island Ecosystems: Biological Organization in Selected Hawaiian Communities.* US/IBP Synthesis Series 15. Hutchinson Ross Publishing Company. Woods Hole, Massachusetts. 583 pp.

Mueller-Dombois, D. and F. R. Fosberg. 1998. *Vegetation of the Tropical Pacific Islands.* Springer-Verlag. New York. 733 pp.

Mueller-Dombois, D. and L. L. Loope. 1990. Some unique ecological aspects of oceanic island ecosystems. *Monogr. Syst. Bot. Missouri Bot. Gard.* 32: 21–27.

Mueller-Dombois, D. and L. D. Whiteaker. 1990. Plants associated with Myrica faya and two other pioneer trees on a recent volcanic surface in Hawai'i Volcanoes National Park. *Phytocoenologia* 19: 29–41.

Neal, M. C. 1965. *In Gardens of Hawaii*. Bishop Museum Press. Honolulu. 924 pp.

Palmer, D. D. 2003. *Hawai'i's Ferns and Fern Allies*. University of Hawai'i Press. 235 pp.

Porter, J. R. 1972. *Hawaiian Names for Vascular Plants*. College of Trop. Agric., Hawai'i Agric. Expt. Sta., University of Hawai'i Departmental Papers 1. 64 pp.

Smathers, G. A. 1972. *Plant Invasion, Early Succession and Recovery on the 1959 Kīlauea Iki Eruption Site*. Hawaii Volcanoes National Park, Hawaii. Ph. D. Dissertation. University of Hawai'i at Mānoa, Honolulu. 450 pp.

Smathers, G. A. 1981. One hundred years of vegetation succession on a crater floor in Hawaii. Pages 61–76 in: *Second Conference of Scientific Research in the National Parks*. Vol. 2.

Smathers, G. A. and D. Gardner. 1980. Stand analysis of an invading *Myrica faya* population, Hawaii. *Pacific Science* 33: 239–255.

Smathers, G. A. and D. Mueller-Dombois. 1974. *Invasion and Recovery of Vegetation after a Volcanic Eruption in Hawaii*. National Park Service Scientific Monograph Series No. 5. 129 pp.

Stearns, H. T. (ed.). 1985. *Geology of the State of Hawai'i*. Pacific Book Publishers. Palo Alto, California. 335 pp.

Tunison J. T., L. F. Castro, and R. L. Loh. 1998. *Myrica faya tree dieback in Hawai'i Volcanoes National Park and vicinity:*

distribution, demography and associated ecological factors. University of Hawai'i CPSU Tech. Report 122. 24 pp.

Vitousek, P. M. and L.R. Walker. 1989. Biological invasion by *Myrica faya* in Hawaii: plant demography, nitrogen fixation, and ecosystem effects. *Ecological Monographs* 59: 247–265.

Vitousek, P. M., L. R. Walker, L. D. Whiteaker, and P. A. Matson. 1993. Nutrient limitations to plant growth during primary succession in Hawai'i Volcanoes National Park. *Biogeochemistry* 23: 197–215.

Vitousek, P. M., L. R. Walker, L. D. Whiteaker, D. Mueller-Dombois, and P. A. Matson. 1987. Biological invasion of *Myrica faya* alters ecosystem development in Hawai'i. *Science* 238: 802–804.

Wagner, W. L., D. R. Herbst, and S. H. Sohmer. 1990. *Manual of the Flowering Plants of Hawai'i.* Vol. 1 and 2. 1854 pp.

Waite, M. 2006. Checklist of Mosses of Hawai'i Volcanoes National Park. mashuri@hawaii.edu

Walker, L. R. and P. M. Vitousek. 1991. An invader alters germination and growth of a native dominant tree in Hawai'i. *Ecology* 72: 1449–1455.

Wright, R. A. and D. Mueller-Dombois. 1988. Relationships among shrub population structure, species association, seedling root and early volcanic succession, Hawai'i. Pages 87–104 in: *Plant Form and Vegetation Structure.* M. J. E. Werger, P. J. M. van der Aart, H. J. During, and J. T. A. Verboeren (eds.). STB Academic Publishing, The Hague, The Netherlands.

Index

A
ʻAʻaliʻi (*Dodonaea viscosa*), 104
ʻAe fern (*Polypodium pellucidum*)
    ʻ on summit of Puʻu Puaʻi, 36
Amaʻu fern (*Sadleria cyatheoides*) in
    forefront of cinder cone, 64
    in tree mold, 50
    on crater floor, 29
    on southwest slope of cinder cone, 38

B
Bathtub ring, rim around lava lake, 32
Broomsedge (*Andropogon virginicus*)
    along Devastation Trail in Spatter Habitat, 58
    an alien grass, scattered at base of cinder cone, 41
    first alien invader in Kāʻū Desert, 105-107
    on cinder-ash compacted by human trampling, 72
    on unshaded side of successional ecotones, 75
Butterfly bush (*Buddleia asiatica*), huelo ʻīlio
    across boardwalk (Devastation Trail), 48
    a transient invader, 51
    invading a dead kūpaoa mat, 65
    rapid advance of alien shrub, 47

C
CESU, Cooperative Ecosystem Studies Unit, 7
Cinder-ash
    defined, 6, 59
Colonizer plants
    initial invaders, viii
    native versus non-native, viii
CPSU, Cooperative Park Studies Unit, 7
Cracks and fissures, 24, 25, 26, 30, 34, 35
Crater floor habitat, 21
    cracks and fissures, 24, 25
    cooling, 29
    Densification in cracks, 28
    first tree seedlings, 25
    Grunow fog interceptor, 24
    habitat constraints, 29
    harsh pioneer habitat, 29
    hot stone desert, 23
    invader sequence, 27
    shrub invaders, 28
    steaming, 29
    sword fern, the false ʻōkupukupu, 27

D
Devastation Area
    an ideal outdoor laboratory, 13
    appearance of alien faya tree, 19
    area extent defined, 6
    board walk in 1960, 45
    first decade study published, 8
    initiation of study, 7, 13
    map, 15
    Nēnē geese, 37, 73
    pioneer vegetation zone, viii
    1250 acres, 500 hectares, 117

Dispersal question, 31

E
Ecosystem
    defined, 12

F
Faya (*Morella faya*), an alien tree
    advanced growth, 93
    engulfing juvenile 'ōhi'a lehua tree, 75
    in desert shrub islands, 105
    nitrogen fixer, 90
    seedling in survival tree habitat, 92
Flora
    defined, 10

H
Habitat and habitats
    brief descriptions with areal extent, 14
    defined, 11
    habitat constraints, 29, 33
    habitat profiles, 16, 17
    six habitats (see figure 2, 15)
Halema'uma'u crater, 4, 15, 98
Hawai'i Natural History Association, 7
Hanu-paoa (*Dubautia ciliolata*), 105

I
IBP, International Biological Program, 7
Initial invaders
    colonizer plants, viii
    false 'ōkupukupu, 27
    huelo 'īlio or butterfly bush, 47, 48, 51
    kūpaoa, 35, 41, 60, 61
    'ohelo 'ai, 26, 30, 34, 37
    'ōhi'a lehua, 25, 26, 29, 32, 40, 41, 42, 43
    thimbleberry, 47
Interaction summary, 122-123
in situ succession study, vii
Invader sequence, 18

K
Ka'ū Desert
    adventitious rooting, 99
    cementation beneath new surface, 99
    desert concept defined, 97
    desert wadis, 106
    first alien invader, 105
    habitat 6, 16
    new opportunity, 98
    shifting ash, 106
    shrub islands, 101-102
Keanakāko'i crater, 4, 15, 108-112
    lava floor surface of 1877 in 1967, 109
    new surface in 1974 as seen in 2002, 112
Kīlauea caldera, 4, 15, 17
Kīlauea Iki crater
    earlier lava flow in 1868, 21
    eruption, 1
    height of fountain, 1
    hot stone desert, 23
    lava lake depth, 1
    lava temperature, 1
    location, 3, 4
    plant invasion summarized, 118
    profile, 16
    view of fountain, 2
Knotweed (*Persicaria capitata*)
    becoming very evident in early 1980s, 66
    invading dead Kūpaoa mat, 63
    in successional ecotone, 74

surrounding a broomsedge bunchgrass, 77
Koa (*Acacia koa*)
    native invader, 91
    nitrogen fixer, 91, 94
Kūkaenēnē (*Coprosma ernodeoides*)
    a native mat former in life form similar to the alien knotweed, 66
    developing around bases of trees, 90, 96
Kūpaoa (*Dubautia scabra*) cushion shrub
    facilitator of subsequent invaders, 63
    seedling in summit community, 5
    senescing mats, 63
    short life cycle, 60
    the first significant colonizer in snag habitat, 60, 61
    with white flowers on south slope, 41

L

Landscape defined, 12
Lava trees, 113
Lichens
    Stereocaulon on cinder cone, 36
    Cladonia on fallen snags, 69, 70
LTER, long-term ecological research, v, 7

M

MAB, Man and the Biosphere, 8, 9
    designation in 1980, 8
    IBP connection, 8
    view of plaque, 8
Montane rain forest, 17
    boundary with seasonal forest, 15
    environment, 14
Montane seasonal forest
    boundary with rain forest, 15
    environment, 14

N

Nēnē (*Branta sandvicensis*) native goose
    a flock at base of Puʻu puaʻi cinder cone, 37
    walking from cinder cone, 73
    well camouflaged in lower snag habitat, 72

O

ʻŌhelo ʻai (*Vaccinium reticulatum*) shrub
    in summit joint crack, 34
    near center of lava lake, 30
    on slope of cinder cone, 37
    seedling in lava crack, 26
ʻŌhiʻa lehua (*Metrosideros polymorpha*)
    advanced seedlings on east slope, 41
    in crack on south slope, 40
    in summit joint crack, 34, 35, 43
    invasion on lower east slope of cinder cone, 42
    juveniles in center on lava lake, 29, 31
    on "bathtub ring", 32
    sapling in flower, 26
    seedling in crevice on lava lake, 25
    some snags become alive, 53, 68
    survivors, 115-116
    tree snags, 51

P

Pāhoehoe lava rock, 32
    shelly lava, 113
Park personnel acknowledgements, ix

Naturalists, ix
Rangers, ix
Superintendents, ix
Park Research Center
    Agencies working at Kīlauea Field Station, 10
    Kīlauea Field Station View, 9
Pāwale (*Rumex skottsbergii*)
    a less common native shrub invader, 65
    dead bush (short life cycle), 116
    engulfing dying huelo 'īlio bush, 66
PCSU, Pacific Cooperative Park Studies Unit, 7
Pioneer vegetation zone
    Devastation area, viii, 6
Pioneer communities, 32
    community formation, 36
    kūpaoa cohort colony, 61
    'ōhelo 'ai shrub community, 39
    summit community, 34, 36, 43
    vegetation island community, 73
Plot and transect surveys, 18
Primary succession
    200 yr to mature rain forest, viii
Pūkiawe (*Leptecophylla tameiameiae*)
    on southwest slope of cinder cone, 38
Pu'u Pua'i cinder cone "Hill of high fountain", 33
    development of summit community, 34
    east slope, 41
    first view, 5
    habitat constraints, 33
    height, 5
    in early stage of invasion, 38
    invasion summary, 43, 119-120
    profile diagram, 16, 17
    slope community, 39
    south slopes, 40, 41
    summit community, 43
    summit joint crack, 34
Pyroclastic tephra
    defined, 6

R

Rain forest, 17, 21, 33
Recovery of vegetation,
    in habitat 3 (spatter habitat) at Devastation Trail, 53
    'ōhi'a lehua snags recover in snag habitat, 67
    recovered snag in flower, 68
    view of survival tree habitat 5 as recovered in 2006, 96
    view of survival tree habitat 5 prior to recovery, 79
    view of spatter habitat 3 prior to recovery at boardwalk (later paved), 45
Reviewers acknowledged, x

S

Seasonally dry forest, 99, 79
Sequence of invaders, 18
Shrub islands, 101
    shrub island associates, 103
Snag habitat
    short-lived perennial plants, 121
    vegetation dynamics during early succession, 120
Sinkholes, 62, 71
Space for time substitution, vii
Spatter habitat 3, 45
    after installation of boardwalk, 45
    Devastation Trail, 46-48
    favorable microhabitats, 48
    initial invasion, 46
    invading front of alien thimbleberry, 47

INDEX

recovery of vegetation in 2006, 58
spatter defined, 45
Specialists acknowledged, ix
bluegreen algae, ix
bryophytes, ix
lichens, ix
Succession
early succession, 56
primary succession, viii
successional ecotones, 74-75
time window, viii
Succession, types studies, vii
in situ succession studies, vii
space-for-time substitution studies, vii
Survival tree habitat
adventitious roots, 84, 100
aerial roots, 83, 84
after deposit of ash blanket in 1960, 79
alien herbaceous invaders, 85-87
declared a ""Special Ecological Area", 90
fertilization effect, 81, 82
habitat defined, 79-80
juvenile tree forms, 80, 95
life returned quickly, 80
as recovered in 2006, 96
shrub invaders, 88
tree invaders, 90
Sword fern, false ʻōkupukupu (*Nephrolepis multiflora*)
at base of tree snag in spatter habitat 3, 49
in cinder-ash depression, 70
plant life invades new rock, 24, 26
sword fern becomes abundant, 27

T

Technical assistance acknowledged, ix
periodic resurveys, ix
special subprojects, ix
Transect system, 114
Trade wind (northeast)
building cinder cone, 22
dry wind with desiccating effect, 98
spreading ash blanket, 22
Tree snag habitat
cinder-ash defined, 59
snags become alive, 66, 67
the first significant colonizer, 60, 61

U

ʻUki (*Machaerina angustifolia*), 106
ʻUlei shrub (*Osteomeles anthyllidifolia*), 104
Uluhe fern (*Dicranopteris linearis*)
in deeper fissures of cinder cone summit, 34
in successional ecotone, 74

V

Vegetation
defined, 10
development defined, 11
surveys, 18

W

World Heritage Site
designation in 1987, 10
IBP connection, 8
view of plaque, 9

About the Authors

Dieter Mueller-Dombois, Ph.D., Dr. h.c.

Born and raised in Germany, Dr. Dieter Mueller-Dombois has traveled the world while teaching and working in the fields of conservation and botany for over fifty years. Considered the foremost vegetation ecologist in the Pacific, Dr. Mueller-Dombois is a professor emeritus at the University of Hawai'i and successfully instituted a world-class plant ecology program during his tenure. His publications include over 200 scientific articles, books, and textbooks such as *Aims & Methods of Vegetation Ecology, Island Ecosystems: Biological Organization in Hawaiian Communities,* and *Vegetation of the Tropical Pacific Islands.* He has also produced an online manual for interactive ecology and management about tropical island ecosystems that can be found at *www.botany.hawaii.edu/pabitra.* At eighty, Dr. Mueller-Dombois continues to travel to Fiji, Japan, Europe, New Zealand, and other Pacific islands to research, lecture, and participate in the scientific communities of these countries. He has lived with his family in the islands since 1963, currently residing in Kailua.

Garrett A. Smathers, Ph.D.

Holding multiple degrees in chemistry, bioecology, psychology, education, and botanical sciences, Dr. Smathers has enjoyed a 27-year career in the National Park Service as a naturalist, chief scientist, program director, teacher, and consultant. He has also worked at the Hawai'i Volcanoes National Park, numerous National Park Service areas, and the NASA Space Technology Laboratories. He is committed to promoting conservation and environmental education. At eighty, Dr. Smathers, now retired, continues his research efforts in ecological studies in Hawai'i and the Southern Appalachians. Dr. Smathers has produced a diverse body of scientific articles, research-resources management plans and reports. He currently lives in western North Carolina and continues environmental consulting work.

From left to right: Garrett A. Smathers and Dieter Mueller-Dombois